Improving DoD's Weapon System Support Program

A Critical Readiness Driver Approach

Marc Robbins, James R. Broyles, Josh Girardini,
Kristin Van Abel, Patricia Boren

Prepared for the Office of the Secretary of Defense
Approved for public release; distribution unlimited

For more information on this publication, visit www.rand.org/t/RR2496

Library of Congress Cataloging-in-Publication Data is available for this publication.
ISBN: 978-1-9774-0157-1

Published by the RAND Corporation, Santa Monica, Calif.

Cover: U.S. Air Force photo by MSgt Stefan Alford.

Preface

Providing for the readiness and sustainability of the armed services' weapon systems is one of the key responsibilities of the Department of Defense's supply chain. A critical component of that supply chain is the vast array of consumable repair parts needed for weapon system maintenance at the field and depot levels. The Defense Logistics Agency (DLA) has responsibility for managing those consumable repair parts (meaning parts that are used one time and do not undergo repair). In 1981 the Weapon System Support Program (WSSP) was established to help DLA and its customers prioritize among this vast array of items. However, WSSP performance has fallen short of its goals, and this shortfall has persisted over decades. A primary reason for the shortfall is the failure to use tools and methodologies for accurately determining true critical weapon system readiness drivers (RDs) and differentiating these items from the rest of the population.

In order to help address and correct the shortfalls of the WSSP, the Deputy Assistant Secretary of Defense for Supply Chain Integration, under the Assistant Secretary of Defense for Logistics and Materiel Readiness, asked RAND to analyze the problems limiting WSSP effectiveness and offer recommendations for improving the program's ability to provide weapon system readiness support. This report documents the results of that study.

In conducting the study, we used a combination of data analysis and interviews with armed services subject matter experts to identify factors underlying the WSSP's performance shortfalls. Using this information and leveraging earlier work, such as RAND tools currently driving retail Army stockage decisions (Girardini, Lackey, and Peltz, 2007), we developed an alternative method for targeting RDs for enhanced support. As a proof of concept, we developed a simulation showing the potential trade-off between readiness driver/non-readiness driver (NRD) investment allocation and materiel availability for Army and Air Force historical deadlining demands.

This research was sponsored by the Office of the Undersecretary of Defense for Logistics and Materiel Readiness (Supply Chain Integration) and conducted within the Acquisition and Technology Policy Center of the RAND National Defense Research Institute, a federally funded research and development center sponsored by the Office of the Secretary of Defense (OSD), the Joint Staff, the Unified Combatant Commands, the U.S. Navy, the U.S. Marine Corps, the defense agencies, and the defense intelligence community.

For more information on the RAND Acquisition and Technology Policy Center, see www.rand.org/nsrd/ndri/centers/atp or contact the director (contact information is provided on the webpage).

Contents

Preface iii
Figures and Tables vii
Summary ix
Acknowledgments xix
Abbreviations xxi

CHAPTER ONE
Introduction 1
Outline of the Document 3

CHAPTER TWO
An Overview of the Weapon System Support Program 5
Background and Overview 5
History and Growth of the WSSP 7
Problems in WSSP Performance Historically 8
Current Issues with WSSP Performance 10

CHAPTER THREE
An Alternative WSSP Approach 21
An Overview of the RAND Approach 21
True Readiness Drivers 21
Reducing the WSSP NIIN Population by Using Field Maintenance–Based Critical RDs 28

CHAPTER FOUR
Modeling Alternative Approaches: An Army/Air Force Case Study ... 33
Overview ... 33
General Statistics on the Demand Population ... 34
The Simulation Approach ... 36

CHAPTER FIVE
Results of the Army/Air Force Case Study ... 45
Reallocating OA from the Investment Base to RDs ... 45
Alternative Investment Strategies ... 46
Results ... 49

CHAPTER SIX
Toward an Integrated DoD Approach ... 53
Moving Toward a Revamped WSSP Using Critical Readiness Drivers ... 53
From Proof of Concept to an Implemented System ... 55

CHAPTER SEVEN
Summary, Conclusions, and Recommendations ... 57
Summary and Conclusions ... 57
Recommendations ... 59

APPENDIX
The Simulation Model ... 61

References ... 69

Figures and Tables

Figures

S.1 OA for Noncritical NIINs by Unit Price Group and Investment Alternative ... xiii
S.2 OA for Critical RDs by Unit Price Group and Investment Alternative ... xiv
S.3 Aggregate MA by NIIN Type Across the Three Cases ... xv
S.4 Net Change in MA by Price Group: Base Case and 20% Reinvestment Case ... xv
2.1 MA, All DLA-Direct Issues, 2016 ... 19
3.1 Most M1 Abrams Tank WSSP EC 1 NIINs Are Not RDs ... 26
3.2 One-Fourth of Abrams Tank RD NIINs Have High WSSP Priority ... 27
4.1 Two-Year Demand Value for Critical and Noncritical NIINs, by Unit Price Group ... 35
4.2 Actual MA for DLA Issues, 2014–2015: RD and NRD NIINs ... 40
4.3 Simulated MA for DLA Issues, 2014–2015: RD and NRD NIINs ... 41
5.1 Yearly Demand Value for Air Force/Army-Dominated NIINs and Critical Readiness Drivers ... 46
5.2 Reallocation of OA to RDs ... 47
5.3 OA for NRDs by Unit Price Group and Investment Alternative ... 48
5.4 OA for RDs by Unit Price Group and Investment Alternative ... 49
5.5 Aggregate MA, by NIIN Type Across the Three Cases ... 50
5.6 RD MA Increases, by Unit Price Group ... 51
5.7 Net Change in MA, by Price Group: Base Case, and 20 Percent Reinvestment Case ... 52

Tables

2.1 Essentiality Codes ... 6
2.2 The WSSP NIIN Prioritization Matrix ... 7
2.3 Systems' WSGC Assignments, by Service, 2014 and 2016 ... 12
2.4 WSGC Changes (A to C, or C to A), by WSDC, 2014 and 2016 ... 13
2.5 WSSP NIIN Population, by Priority, 2016 ... 15
2.6 DLA Issues by NIIN Priority, All DLA-Direct Issues, 2016 ... 16
2.7 Value of WSSP NIIN Demand, 2014–2015, by Priority ... 17
2.8 Value of Demand for WSSP High-Priority NIIN, by Service Designating Priority, 2014–2015 ... 18
3.1 Air Force MICAP NIIN Actions, 2011–2016 ... 28
3.2 Comparison of Air Force High-Priority WSSP NIINs and MICAP-Based Critical RDs ... 30
3.3 Comparison of Army High-Priority WSSP NIINs and MICAP-Based "True" RDs ... 31
4.1 Number of and Demand for Army/Air Force RDs and NRDs, and Distribution of Unit Prices, 2014–2015 ... 34
4.2 Aggregate Demand for Army/Air Force RDs and NRDs, 2014–2015 ... 35

Summary

Providing for the readiness and sustainability of the armed services' weapon systems is a key responsibility of the Department of Defense's (DoD's) supply chain. A critical element of that supply chain is the vast array of consumable repair parts needed for weapon system maintenance at the field and depot levels. The Defense Logistics Agency (DLA) has responsibility for managing those consumable repair parts (meaning parts that are used one time and do not undergo repair).

The Weapon System Support Program (WSSP) was established to help DLA and its customers prioritize among this vast array of items. Its intent was to increase collaboration between customers and their major suppliers by which the services would identify items of greatest priority to them, and DLA would focus its management attention and limited resources on prioritizing the availability of those items.

However, WSSP performance has fallen short of its goals for decades. A primary reason for this shortfall has been the failure to use existing tools and methodologies for accurately determining true critical weapon system readiness drivers (RDs) and differentiating them from the rest of the population.

To help address and correct the shortfalls of the WSSP, the Deputy Assistant Secretary of Defense for Supply Chain Integration, under the Assistant Secretary of Defense for Logistics and Materiel Readiness, asked RAND to analyze the problems limiting WSSP effectiveness and offer recommendations for improving its ability to provide weapon system readiness support. This report documents the results of that study.

Specifically, this report answers the following research questions:

1. How can the WSSP be used to allow DLA to improve support for service-prioritized weapon systems?
2. Based on existing Air Force and Army evidence, how could all services develop effective prioritization inputs to the WSSP, and what should a common approach across the services look like?
3. How much could support for RDs be increased, and what would be the consequences for the remaining non-readiness driver (NRD) population in a fixed-cost environment?

Current WSSP Issues

Previous independent assessments have identified problems with WSSP performance, and this research effort has identified several issues that have persisted for decades:

1. The number of WSSP National Item Identification Numbers (NIINs) is large, approaching 2.5 million.[1] Almost one-third of this number are at the highest level of priority, and one-half are in the top five tiers of priority (out of 15 tiers).
2. In addition, a preponderance of WSSP NIINs are in the highest-priority WSSP categories. The highest-priority NIINs account for 60 percent of all DLA issues. Moreover, those demands are disproportionate across the services, with the Navy high-priority NIINs alone accounting for almost 40 percent of total value of all materiel DLA issues.
3. Partly due to this, the system delivers materiel availability (MA) no higher for high-priority NIINs than for the general population.[2]

[1] The National Item Identification Number is a unique nine-digit identifier code for a specific type of part.

[2] Materiel availability is a measure that indicates whether a needed item is available for immediate shipment or is instead back-ordered and awaiting replenishment. DLA's goal is to achieve an MA of at least 90 percent.

These issues pose challenges for directing investment in the WSSP and contribute to the shortfalls in meeting the program's goals of producing higher availability for the most critical items.

Alternative Approaches: An Army/Air Force Case Study

We developed an alternative WSSP approach focused on identifying critical RDs as a means of prioritizing among items. Critical RDs are those NIIN components that are especially crucial in weapon system "deadlining" actions (i.e., when the system is not mission capable); while failures of many NIINs can and do deadline weapon systems, we define the most critical ones as those that dominate the population of deadlining events. Using field-level maintenance data for the Army and Air Force, our approach sought out the smallest number of NIINs that made up at least 50 percent of all deadlining actions for the Weapon System Group Code (WSGC) A and B systems, which make up the higher priorities in the WSSP.[3] In addition, each NIIN so identified had to contribute a sizable number of deadlining actions for the weapon system it was associated with. Lastly, the number of deadlining actions had to exceed a minimum percentage of total requisitions made by the service for the particular NIIN.

Using these criteria, we developed a critical RD list that is a much smaller portion of the NIIN population coded as being of high priority in the WSSP. For example, the WSSP NIIN count for the highest-priority items for WSGC A systems for the Army is 78,622. We identified 9,778 critical RD NIINs for that same group of systems, about 13 percent of the WSSP NIIN population. Similarly, for the Air Force, the WSSP NIIN count for the highest-priority items for WSGC A and B systems is 429,630. We identified 4,833 as being critical, which amounts to only about 1 percent of the WSSP NIIN population.

---

[3] Three data sources were used: the Equipment Downtime Analyzer (EDA), the Army Aviation Aircraft on Ground (AOG) database, and the Air Force Logistics, Installations and Mission Support–Enterprise View (LIMS–EV). This study was limited to Army and Air Force data due to constraints in the research agenda. Similar work for the Marine Corps and Navy is still yet to be done to work toward an integrated DOD-wide approach.

Redirecting investment to these critical RDs should have the effect of increasing MA for these items and improving overall performance of the WSSP.

Simulation: Reallocating OA from the Investment Base to RDs

To conduct a proof of concept, we developed a simulation to show the trade-off between RD and NRD investment allocation and MA. The simulation did not strive to replicate DLA's procurement logic, nor did it intend to forecast MA as a function of investment allocation; rather, the intent was to replicate current performance as a way to explore the potential for MA versus investment allocation trade-offs as a precursor to exploring real-world investment reallocation and MA effects.

The simulation mimics observed MA by unit price group and then models hypothetical investment scenarios, where available funds for NRD replenishment are reallocated so as to increase the safety stock of RDs. This has the effect of increasing MA for the RDs and decreasing MA for NRDs. We express the reallocation value as a percentage of the annual RD demand.

We examine three cases for investing more in critical RDs:

- The base case: the current investment approach, with no preference given to critical RDs.
- A one-time increase in investment in critical RD safety stock, equal to 20 percent of the yearly demand for RDs.
- A one-time increase in investment in critical RD safety stock, equal to 40 percent of their yearly demand.

In the latter two cases, funds for higher RD safety stocks would come from reducing investments in NRDs, in effect lowering their safety stock levels, with concomitant declines in their expected MA.

Results

Our simulation optimizes reallocation of a fixed level of Obligation Authority (OA) to achieve maximum overall MA given historical variability of demand. In the simulation, most investment goes to the more expensive items, primarily because most demand value derives from those NIINs. In general, then, we see substantial reallocation of OA from expensive NRD NIINs to the more expensive RD items. Total OA is dominated by NIINs with unit prices between $1,000 and $10,000. RDs in that range receive the biggest increase in investment; the same group of noncritical NIINs show the largest decrease in replenishment investment in the 20 percent and 40 percent cases compared to the base case (see Figures S.1 and S.2).

Because this investment strategy is not focused on aggregate optimized MA, overall MA drops. That is, MA for noncritical items drops to a greater extent than it correspondingly increases for critical items (see Figure S.3).

Figure S.1
OA for Noncritical NIINs by Unit Price Group and Investment Alternative

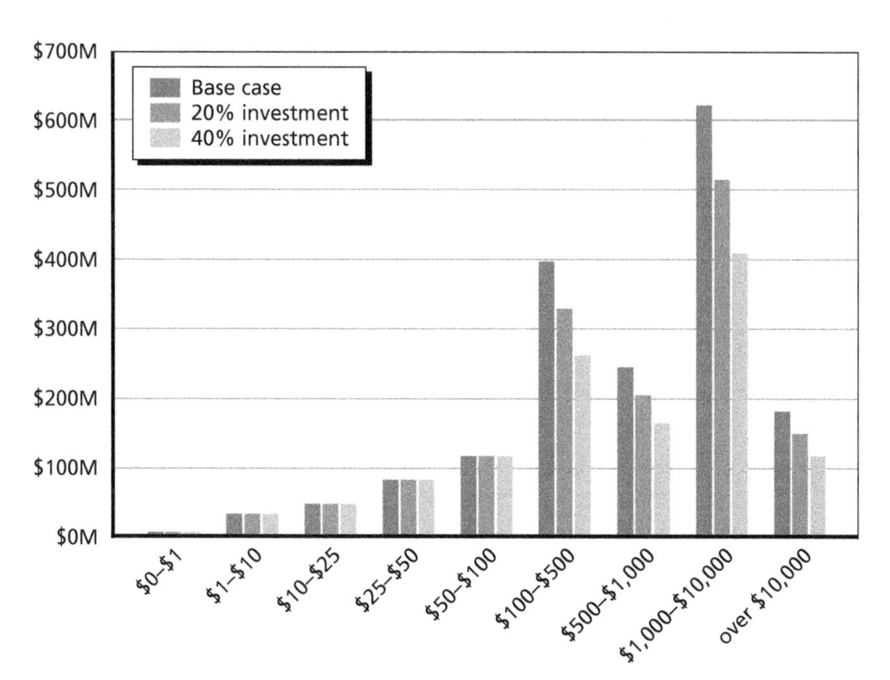

NOTE: OA for NIINs with unit price below $100 protected.
RAND *RR2496OSD-S.1*

Figure S.2
OA for Critical RDs by Unit Price Group and Investment Alternative

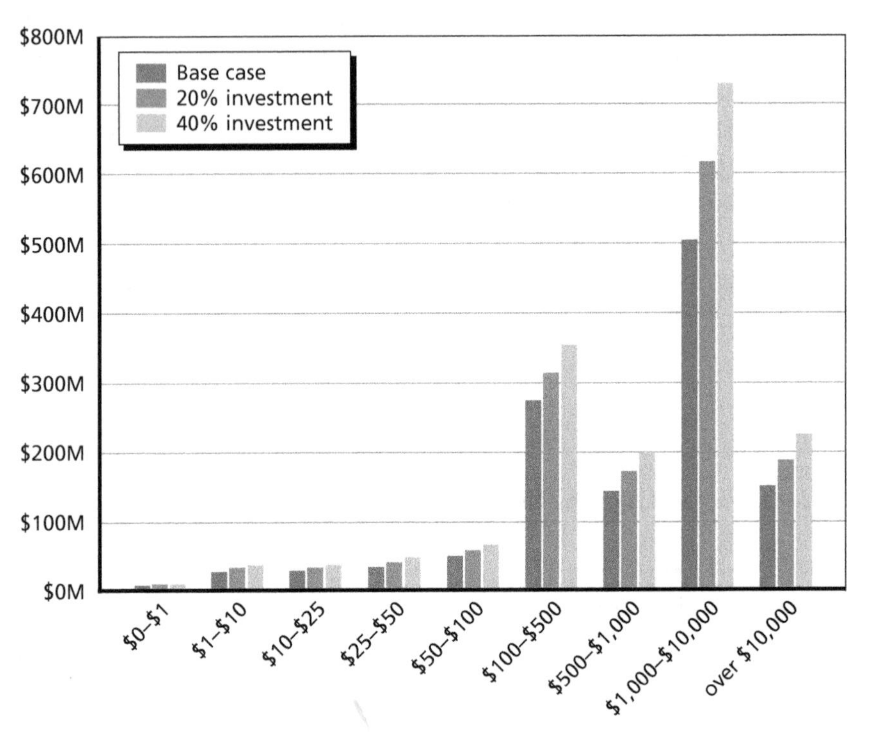

RAND *RR2496OSD-S.2*

Under the two alternative reinvestment approaches, total and NRD MA fall compared with the base case, while MA for RDs increases. Most of that increase goes to the more expensive RD parts, which have traditionally been relatively underresourced and thus have tended to have the lowest MA. Reallocating OA can yield substantial benefits for this group, while the MA for the more expensive NRDs is allowed to fall, as Figure S.4 illustrates for the 20 percent reinvestment case.

Figure S.3
Aggregate MA by NIIN Type Across the Three Cases

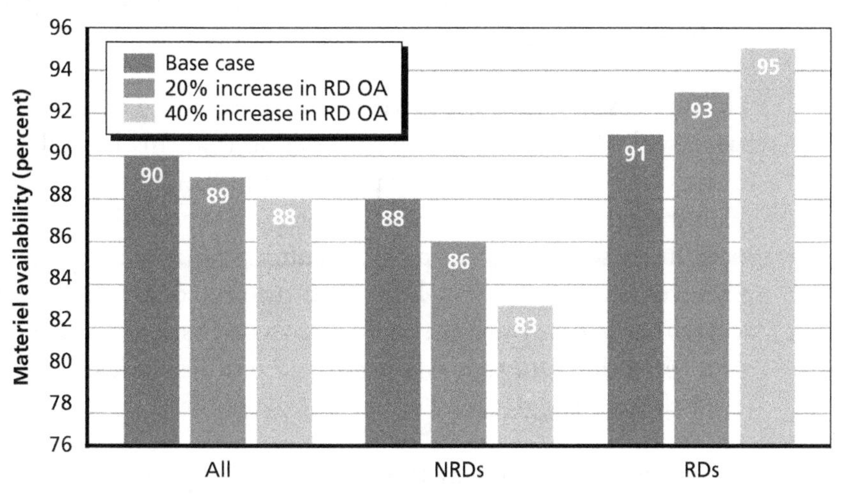

RAND *RR2496OSD-S.3*

Figure S.4
Net Change in MA by Price Group: Base Case and 20% Reinvestment Case

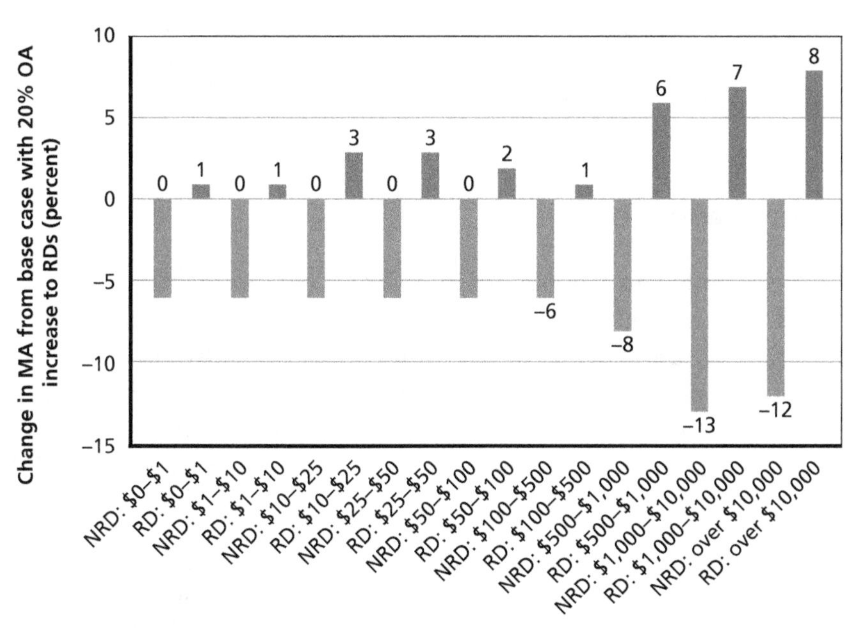

RAND *RR2496OSD-S.4*

Toward an Integrated DoD Approach

Making the WSSP a program that successfully supports critical weapon system readiness, as argued in this report, requires two actions:

1. The services should revise methods for determining critical RD parts, with the almost certain implication that this list will be far smaller than that currently provided to DLA.[4]
2. DLA should adopt a procurement strategy that prioritizes investment in these critical RD parts. Assuming no new investment resources, that means that DLA (and the armed services) must accept that *lower-priority* items will necessarily see lower MA. It also means that DoD as a whole must accept "less efficient" MA results from an aggregate, enterprise-wide point of view.

Developing agreed-upon methods for generating lists of critical RDs is key to improving the WSSP and building consensus for its outcomes. These new lists may take time, but progress toward improving the WSSP does not necessarily require complete and final development across all four services. Partial, phased implementation is possible. While the services are ultimately responsible for producing these new approaches, this report presents examples of how that might be done and what challenges will be involved.

Recommendations

As documented in this report, the WSSP has not achieved its goal of prioritizing support for high-priority weapon systems. The primary reason, as argued here, is the lack of the services' ability either to identify or communicate to DLA the relatively small portion of the repair part population that dominates readiness problems for their equipment. Instead, high-priority parts lists shared by the services with DLA

[4] The methodologies used must be transparent, understandable, and defensible to all parties in the program, and the services need to make efforts to avoid even the appearance of "gaming the system" for a new critical RD-focused investment strategy to be effective.

tend to be inflated and, indeed, account for a large majority of *everything* DLA stores and issues. It is no surprise, then, that WSSP items, even those designated as being of high priority, are typically treated no differently from the run-of-the-mill materiel that DLA manages and generally show the same MA as anything else.

But the fact that the WSSP has not delivered higher MA for critical RDs does not mean that it cannot. The Army already uses a critical RD approach to guide inventory stockage at its brigade-level supply support activities. The same logic, this report argues, can and should be used to guide a new approach to supporting the WSSP.

RAND offers the following recommendations for improving the WSSP:

1. **The Office of the Secretary of Defense (OSD) should develop and promulgate policy justifying and providing for a new approach to the WSSP.** This policy should seek to reduce the overall population of high-priority NIINs in the WSSP, with means to be determined by each service, and provide a justification for doing so; it should also set the terms for balancing the size of those populations across the services in order to achieve fair allocation of resources across all services.
2. **The services should revise their approaches for determining critical RDs and present justifications for those approaches and the resulting lists.** Using field maintenance data, as described in this report, would be one way of identifying critical RDs. While the ultimate determination will be made by the services, they need to be aware that equitable allocation of resources will imply constraints on the size of those lists and could recast OA investment.
3. **The services should seek to make minimal changes in their assignment of system-level priority (as defined by the WSGC) to their candidate systems, and should review and revise their critical RD NIIN lists according to a regular schedule.** Investments to achieve higher MA may involve long lead times and substantial sunk costs. It is, therefore, important for the services to limit volatility in WSGC assignments and the tur-

bulence such volatility creates in critical RD lists. At the same time, the services must ensure that critical RD lists stay as up to date as possible; this means both adding NIINs that are becoming increasingly important for readiness (e.g., due to aging) and removing those that are no longer critical.

4. **DLA should strive to achieve higher MA for critical RD safety stock by pursuing policies and procedures that prioritize OA investments in those items.** DLA should set target MA levels for the revised WSSP critical RD lists and allocate OA in order to achieve those targets. Because aggregate MA is heavily influenced by the cheapest items (even if those items are not always the most important to driving readiness), DLA may need to establish explicit higher targets for more expensive NIINs, beyond what would be achieved through standard inventory management approaches.
5. **DLA should report MA results disaggregated into RD and NRD populations, as well as overall MA. Additionally, after advisement from OSD (and other stakeholders), DLA should pursue an approach that lowers overall MA in exchange for higher RD MA.** OSD policy and required metrics, in particular, should reflect the revised investment strategy, and help support the disaggregated investment approach that DLA would adopt under a new WSSP.

Acknowledgments

This study would not have been initiated nor brought to a conclusion absent the interest and strong support from our sponsor, Deline Reardon, the Deputy Assistant Secretary of Defense for Supply Chain Management. We thank her for that support and that of her exceptional staff members, Paul Blackwell, Bob Carroll, and Jan Mulligan.

We appreciate the help with WSSP-related data we received from the DLA Office of Operations Research and Resource Analysis, and thank its director, Ken Mitchell, as well as Randy Wendell.

The project received critical feedback from Headquarters, DLA, and we thank Mike Scott, the deputy at Headquarters, DLA J3, along with Mark Melius, Elizabeth Riley, and Emily Vogeler.

The project team was fortunate to interact with senior leaders and subject matter experts among the services. In particular, we thank Peter Bechtel, the Director of Supply Policy (Headquarters, Department of the Army G4) and members of his staff, Paul Hays and Dave Irvin. We appreciated the opportunity to sit down with MG Claude LeMasters, commanding general of Army Tank-Automotive and Armaments Command. We also thank Brent Swart, of the Army Materiel Command and Aviation and Missile Command for his assistance and expertise. RADM James Stamatopolous (Office of the Chief of Naval Operations N41), kindly provided an opportunity to provide a Navy perspective. Our Air Force analysis benefited from the insights of David Wright (Headquarters, Air Force Materiel Command), CMSgt Leighton Sinclair (23rd Logistics Readiness Squadron), and RAND Project AIR FORCE fellow Lt Col Brian Ballew.

Cynthia Cook, former director of the Acquisition and Technology Policy Center of the RAND National Defense Research Institute, provided support and guidance throughout the course of this research. We also owe a debt of gratitude to other RAND colleagues, including Gordon Lee for attempting to whip the document into better shape, and Ken Girardini and Pat Mills for providing their insights and help. Lastly, we thank Pam Thompson for her able management of the document.

Abbreviations

AOG	Aircraft on Ground
DLA	Defense Logistics Agency
DoD	Department of Defense
DORRA	DLA Office of Operations Research and Resource Analysis
EC	Essentiality Code
EDA	Equipment Downtime Analyzer
IP	inventory position
LIMS–EV	Logistics, Installations and Mission Support–Enterprise View
MA	Materiel Availability
MICAP	Mission Impaired Capability Awaiting Parts
NIIN	National Item Identification Number
NRD	non-readiness driver
OA	Obligation Authority
OSD	Office of the Secretary of Defense
PALT	Production and Administrative Lead Time
RD	readiness driver
RO	Requisition Objective
SDDB	Strategic Distribution Database
UJC	Urgency Justification Code

WSDC	Weapon System Designator Code
WSGC	Weapon System Group Code
WSIC	Weapon System Indicator Code
WSSP	Weapon System Support Program

CHAPTER ONE
Introduction

Providing for the readiness and sustainability of the armed services' weapon systems is one of the key responsibilities of the Department of Defense's (DoD's) supply chain. A critical component of that supply chain is the vast array of consumable repair parts needed for weapon system maintenance at the field and depot levels. The Defense Logistics Agency (DLA) has responsibility for managing those consumable repair parts (meaning parts that are used one time and do not undergo repair). It must find and manage manufacturers or vendors, write and execute contracts for replenishment, determine and maintain the appropriate stock levels to meet service level objectives for having needed parts available when requested, and do so at the least cost possible (e.g., avoiding both under- and overbuying stocks of these parts).

In 2014–2015 DLA issued $11.1 billion of materiel to over 21,000 customers of the four services, across some 540,000 different items.[1] It is charged to maximize the chances that a part ordered will be immediately available for issue, with an aggregate goal of over 90 percent materiel availability (MA)—that is, items for customer requests should be

[1] Customers are defined in terms of unique DoD Activity Address Codes, which identify a DoD unit, activity, or organization that has the authority to requisition, contract for, receive, have custody of, issue, or ship DoD assets, or fund/pay bills for materials and/or services. This management of materiel applies to DLA-owned stock issued from DLA-operated depots in the United States and overseas only. DLA also manages programs based on direct delivery from vendors to DoD customers, accounting for roughly another 10 percent of customer demand for DLA-managed items. We do not include this activity in our study because it does not fall under the aegis of the WSSP, and DLA is not responsible for buying the materiel or managing stock levels, which remains the responsibility of the vendor.

back-ordered no more than 10 percent of the time.[2] The large scale of DLA's supply chain responsibilities provides a context for the management challenge DLA and the services face in deciding how to allocate limited resources to prioritize availability of those items most critical to weapon system readiness. But which aspect of that two-year demand of over $11 billion should get preference, and which of those 540,000 different items?

The Weapon System Support Program (WSSP) was established to help DLA and its customers prioritize among this vast array of items. Its intent was to increase collaboration between customers and their major suppliers such that the services would identify items of greatest priority to them and DLA would focus management attention and its limited resources on maximizing those items' availability.[3]

However, as this report demonstrates, WSSP performance has fallen short of its goals, and that shortfall has persisted over decades. A primary reason for the shortfall has been the failure to use tools and methodologies for accurately determining the true critical weapon system readiness drivers (RDs) and differentiating them from the rest of the population. Rectifying this would pave the way for more effective targeting of DLA's limited resources, and should, if implemented, yield significant increases in availability of these critical items.

In order to help address and correct the shortfalls of the WSSP, the Deputy Assistant Secretary of Defense for Supply Chain Integration, under the Assistant Secretary of Defense for Logistics and Materiel Readiness, asked RAND to perform an analysis of the problems limiting WSSP effectiveness and to offer recommendations for improv-

---

[2] The overall MA goal for DLA-managed hardware items (i.e., regarding such things as repair parts) is currently 91 percent. Performance fell just short of that, at 90.3 percent, in FY 2016; see Defense Logistics Agency (2016, p. 83).

[3] The WSSP is one among a number of agreements between DLA and its customers for achieving support goals given available funds. The WSSP is aimed at wholesale item support of repair parts (i.e., issues from stocks held by DLA at its supply depots to the general population of customers). DLA has separate agreements with the services (currently the Air Force and the Navy) to maintain retail-level (i.e., collocated) supply support at individual service repair depots, with performance targets established by agreement. See Defense Logistics Agency (2015).

ing its ability to provide weapon system readiness support. This report documents the results of that study.

Specifically, this reports answers the following research questions:

- How can the WSSP be used to allow DLA to improve support for service-prioritized weapon systems?
- Based on existing Air Force and Army cases, how could all services develop effective prioritization inputs to the WSSP, and what should a common approach across the services look like?
- How much could support for RDs be increased, and what would be the consequences for the remaining non-readiness driver (NRD) population in a fixed-cost environment?

The research related to the first two questions sets the foundation for improved methods for identifying RDs and improved usage of the WSSP. The research related to the third question shows the potential for increasing support for RDs and provides examples on how to evaluate trade-offs between RD and support not related to readiness. These topics are necessary prerequisites to improving procurement and safety stock logic at DLA.

In conducting this study we used a combination of data analysis and interviews with armed services subject matter experts to identify factors underlying the WSSP's performance shortfalls. Using this information and leveraging earlier work, such as RAND tools currently driving retail Army stockage decisions (Girardini, Lackey, and Peltz, 2007), we developed an alternative method for targeting RDs for enhanced support. As a proof of concept, we developed a simulation showing the potential trade-off between RD and NRD investment.

Outline of the Document

Chapter Two offers a brief overview of the WSSP, discussing its origins and earlier critiques of its performance and offering an analysis of its current performance shortfalls and the main reasons for them. Chapter Three offers an alternative approach, based on defining and identifying critical RDs and illustrating that approach with evidence from the Air Force and Army. Chapter Four lays the basis for an analysis of that alternative approach, proposing a simulation-based case study of the benefits of reallocating investment resources to those critical RDs. Chapter Five presents the results of that analysis, indicating the possible increases in critical RD that could be achieved, again based on an Air Force/Army–focused proof of concept case study. Chapter Six discusses the implications of the proof of concept analysis for DoD supply chain operations, including how it might be expanded to include all customers. Finally, Chapter Seven summarizes the study and offers recommendations for the way forward.

CHAPTER TWO

An Overview of the Weapon System Support Program

Background and Overview

The WSSP was established in 1981 with the objective of enhancing the readiness and sustainability of the military services by providing the maximum practical level of support for DLA-managed items with weapon system application. DLA Regulation 4140.38, *DLA Weapons Systems Support Program* (DLA, 1989) provided policy, established guidance and procedures, and assigned responsibilities for the DLA WSSP. Through the WSSP, DLA aims to develop support strategies geared to weapon system criticality and the essentiality of the DLA parts assigned to them. For the most critical items, DLA uses the WSSP to assign personnel, initiate procurement actions, tailor business arrangements, and focus attention on items that are essential to avoid degradation of mission capability (Headquarters, U.S. Department of the Army, 2009; Office of the DoD Inspector General, 1994; U.S. Department of the Navy, 2011).

WSSP has two major elements: participating weapon systems offered by the services, which fall into three groups by criticality, and a list of DLA-managed component parts (identified by National Item Identification Number, or NIIN) as assigned to each weapon system. Weapon system criticality is identified by an assigned Weapon System Group Code (WSGC). Each service is allowed to assign 30 weapon systems to the highest level of criticality (WSGC A), 50 to the second tier (WSGC B) and as many as it wishes to the lowest level of criticality (WSGC C).

Table 2.1
Essentiality Codes

EC	
1	Failure to this part will render the item inoperable
5	Item does not qualify for the assignment of code 1 but is needed for personal safety
6	Item does not qualify for the assignment of code 1 but is needed for legal, climatic, or other requirements peculiar to the planned operational environment of the end item
7	Item does not qualify for assignment of code 1 but is needed to prevent impairment of or the temporary reduction of operational effectiveness of the end item
3 or blank	Failure to this part will not render the end item inoperable

SOURCE: DoD (2014, Table 3, p. 33).

NIIN essentiality is determined by its Essentiality Code (EC). Independent of the WSSP, ECs are typically assigned by the services during the acquisition phase of weapon system development, though they may be revisited and modified after fielding.[1] There are five EC values, as shown in Table 2.1.

The combination of WSGC and EC yields an overall level of prioritization for WSSP parts defined as the Weapon System Indicator Code (WSIC). Table 2.2 presents the prioritization matrix for WSSP NIINs. The top part of the table shows the assignment of the WSIC as used in the WSSP based on a matrix of weapon system priority (the WSGC) and the NIIN EC. The bottom part of the table shows how those NIINs are ranked in terms of WSSP priority. Note that some, but not all, of the highest-priority weapon systems in WSGC A have high NIIN priority rank; those WSGC A weapon system NIINs with less critical ECs are assigned a lower WSSP NIIN priority.

[1] For background on the definition and policy regarding item essentiality codings, see U. S. Department of Defense (2014).

Table 2.2
The WSSP NIIN Prioritization Matrix

Weapon System Group Code		NIIN EC				
	WSIC	1	5	6	7	3/blank
A		F	G	H	J	K
B		L	M	P	R	S
C		T	W	X	Y	Z
		NIIN Rank				
	NIIN Rank					
A		1	2	3	10	13
B		4	6	8	11	14
C		5	7	9	12	15

SOURCE: Headquarters, U.S. Department of the Army (2009, p. 4).

The highest-priority WSSP NIINs are those with EC 1 for WSGC A systems. The three top-priority NIINs are all for WSGC A systems as well, with the EC 1 systems for lower-priority systems (WSGC B and C) coming in at priority (or Weapon System Essentiality Code) 4 and 5.

History and Growth of the WSSP

After the WSSP's initiation in 1981, participation was modest but grew rapidly as the services increased weapon systems and NIINs identified for WSSP prioritization. In FY 1983 some 475 systems with 748,000 NIINs were listed in the program. Following efforts by DLA to publicize the program, by the early 1990s some 1,222 weapon systems with over 1.3 million NIINs were included in the WSSP (Robinson, 1993, p. 17). This included about 40 percent of the 2.5 million hardware items DLA managed at the time. In 1989 the Air Force had placed more than 350,000 NIINs in the program, the Navy more than 445,000, the Army more than 250,000, and the Marine Corps more than 32,000

(Hanks, 1990, p. 12). By December 1993 DLA managed about 3.5 million national stock numbers, including about 1.9 million national stock numbers in the WSSP identified as supporting 1,403 weapon systems or systems components. Of that 1.9 million, about 762,000 (39 percent) were categorized in the three highest criticality codes (Office of the DoD Inspector General, 1994).[2]

Problems in WSSP Performance Historically

Independent researchers identified challenges with WSSP support in the early 1990s. Maj Nathaniel Robinson of the Air Force presented evidence showing little difference in the MA of NIINs included and not part of the WSSP. (MA, indicates how often a needed item is available for immediate issue—i.e., it does not need to be back-ordered). Between 1987 and 1991 WSSP NIIN availability was almost the same as non-WSSP items, at 89.3 percent MA compared to 88.8 percent for non-WSSP items (Robinson, 1993, Fig. 14, p. 29). Chris Hanks, then a researcher at the Logistics Management Institute, studied how reduced MA for WSSP NIINs contributed directly to decreased weapon system readiness (Hanks, 1990).

In 1994, the Office of the DoD Inspector General issued a report indicating significant problems with WSSP execution and proposing a set of recommendations to improve performance. As the report explained,

> The intent of the WSSP was not being fully achieved because, as of December 1993, the WSSP had grown to a degree where it represented over half the total items managed by DLA. As a result of the size of the WSSP, DLA was unable to manage all items included in the WSSP on an intensified basis, and about

[2] Current NIIN levels in the WSSP are discussed in more detail elsewhere in this chapter. It is worth noting here that the size of the WSSP has increased remarkably in the past several decades. As of 2016 there were 2.9 million service-identified NIINs entered into the WSSP, and 2.4 million unique NIINs (0.5 million were reported by more than one service). The Army accounted for 453,000, the Air Force 773,000, the Marine Corps 175,000, and the Navy 1.47 million NIINs. There are currently 1,916 weapon systems listed in the WSSP.

> 60 percent of the items included in the WSSP received no additional supply support. (Office of the DoD Inspector General, 1994, p. 12)

The report noted that while a goal of the WSSP was to augment investment in critical item inventory stocks in order to provide buffers against variable demand, "the large number of items coded with WSICs of F, L, or T and limited funds presently available results in reduced purchases and defeats the intent of the WSSP, which is to provide maximum support to the Military Departments' most critical weapon systems" (p. 13). The report further noted, "Although the WSSP provides that an extra amount of safety level, or an augmented safety level, may be procured and maintained for WSSP items with WSICs F, L, or T, three of the four DLA hardware centers did not include augmented safety levels in their requirements computations" (p. 13). In light of its findings, the report made three recommendations:

> 1. We recommend that the Commanders, Army Materiel Command, Naval Supply Systems Command, Air Force Materiel Command, the Marine Corps Deputy Chief of Staff for Logistics and the Director, Defense Logistics Agency:
>
> a. Establish formal arrangements for the periodic validation and reconciliation of weapon systems applications files for Defense Logistics Agency managed items.
>
> b. Conduct a joint study to reduce the number of items included in the Weapons Systems Support Program and determine which weapons systems items are to be intensively managed.
>
> 2. We recommend that the Commanders, Army Materiel Command, Naval Supply Systems Command, Air Force Materiel Command, and the Marine Corps Deputy Chief of Staff for Logistics establish the controls necessary to ensure that periodic reviews of weapon systems essentiality codes are performed, as required by DoD Regulation 4140.1R, and that the current status of weapon systems is reflected in the Defense Logistics Agency Weapons Systems Support Program data base.
>
> 3. We recommend that the Director, Defense Logistics Agency, review each of the four hardware centers' supply support policies and develop a consistent policy for supply support of

> weapon systems program items, including essential items classified as nonstocked.[3] (Office of the DoD Inspector General, 1994, p. 18)

Only the Air Force concurred with recommendation 1.b, which called for a joint study to reduce the number of items included in the WSSP. In its response, DLA commented,

> Merely reducing the number of items in the WSSP may exclude items that are readiness drivers and/or maintenance linestoppers. Limiting the population of the WSSP could place DLA at risk of insufficient support for an item that would not have qualified for inclusion in the WSSP based on demand alone. Reconciliation efforts now underway will accomplish the purpose of optimizing resources without diluting the program. (Office of the DoD Inspector General, 1994, p. 54)

Current Issues with WSSP Performance

Volatility in Weapon System Assignment

Supply chain management requires some level of stability in target populations given the often long lead times for executing investment decisions and the risk of "churn" in changing items that are designated as being of high priority. Instability in the WSSP can come from two sources: changing prioritization of weapon systems, and turnover in the item lists associated with those systems. The latter may be a reason for concern, but potentially less than the former. Priority of NIINs is determined by the EC parameter, which is typically set in the acquisition process and not changed afterward (Office of the DoD Inspector General, 1997, p. 2).[4] Changing weapon system priority itself, however,

---

[3] Apart from the Air Force, the services and DLA did not concur with recommendation 1.b to reduce the number of items in the WSSP.

[4] DoD established a uniform essentiality coding structure determined by the degree of essentiality of a part based on the hierarchical relationship of the part to higher assemblies, up to the entire weapon system. ECs are typically determined by engineers or equipment

can throw a wrench into supply chain planning if items associated with a previously high-priority weapon system are downgraded along with that system, or vice versa.

Typically this has not been a problem in the WSSP, but our study found one major exception: great turbulence in Air Force assignments of weapon system priority. The RAND study team examined service weapon system WSGC assignments across two years—2014 and 2016—and looked for changes in WSGCs. Table 2.3 shows how WSGC assignments by the services changed between those two years. It displays the numbers of systems assigned to WSGC A, B, or C in 2014 and their corresponding assignments in 2016. Systems assigned to WSGC A in 2014, for example, could have retained that assignment in 2016 or could have been reassigned to WSGC B or C. A similar outcome could have taken place for the other 2014 assignments.

There was little change for the Army. It assigned 30 systems to WSGC A in 2014; two years later, 28 of those systems were still assigned to WSGC A, whereas two of them had been reassigned to WSGC B. Of the 47 systems that the Army had assigned to WSGC B in 2014, a total of 39 retained that assignment in 2016, with the remaining eight being reassigned to WSGC C.

By contrast, the Air Force redefined a large number of systems, not only moving them to the next adjacent WSGC (e.g., from B to A) but often making more extreme moves. Six weapon systems that were designated WSGC A in 2014 were reassigned as WSGC C in 2016; 13 WSGC C systems in 2014 became WSGC A systems in 2016.[5] The Marine Corps and Navy also made similar moves, but not to the extent the Air Force did.

---

specialists capable of evaluating the criticality of the part for required operations. These determinations are often determined during initial provisioning of weapon systems (i.e., before there has been substantial experience with the system in the field).

[5] It should be noted that some of the volatility observed with Air Force weapon system priority assignments might simply be due to the nature in which the Air Force nominates and manages weapon system designator code (WSDC) and WSGC assignments. According to Air Force Instruction 23-101, new WSDC requests are initially assigned to WSCG C; upgrades to B or A can be made during the next annual review; see U.S. Department of the Air Force (2016).

Table 2.3
Systems' WSGC Assignments, by Service, 2014 and 2016

Service	2014 WSCG	2016 WSCG
Army	A = 30	A = 28 B = 2 C = 0
	B = 47	A = 0 B = 39 C = 8
	C = 592	A = 0 B = 6 C = 586
Air Force	A = 28	A = 13 B = 9 C = 6
	B = 41	A = 4 B = 22 C = 15
	C = 209	A = 13 B = 16 C = 180
Marine Corps	A = 30	A = 24 B = 1 C = 5
	B = 46	A = 1 B = 35 C = 10
	C = 319	A = 1 B = 12 C = 306
Navy	A = 31	A = 20 B = 4 C = 7
	B = 45	A = 1 B = 37 C = 7
	C = 541	A = 0 B = 0 C = 541

Table 2.4 shows the systems that were moved from WSGC C to A or WSGC A to C between the two years.

Table 2.4
WSGC Changes (A to C, or C to A), by WSDC, 2014 and 2016

Air Force			
Weapon System	**WSDC**	**2014 WSGC**	**2016 WSGC**
AIM sidewinder missile systems (9P/L/M/X)	03F	C	A
Support equipment, C-5 aircraft	86F	C	A
Support equipment, E-3a aircraft	95F	C	A
Support equipment, C-135 aircraft	96F	C	A
Support equipment, C-130 aircraft	97F	C	A
Support equipment, B-52 aircraft	ABF	C	A
Aircraft, Osprey CB-22B	CTF	C	A
Support equipment, C-17A aircraft	EDF	C	A
U-2 engine (F118-GE-101)	PBF	C	A
U-2 support equipment	PCF	C	A
F-22 Raptor air dominance fighter	PGF	C	A
Engine, F119-PW-100 (F-22 Raptor)	PHF	C	A
Global Hawk high-altitude-long endurance unmanned aerial vehicle	PQF	C	A
Defense support program	40F	A	C
Helicopter, HH-60	75F	A	C
Engine, aircraft, GE T-700 (UH-60A)	BJF	A	C
Engine, aircraft, F100-PW-200 (F-16A/B/C/D)	BUF	A	C
Engine, aircraft, F100 PW-220 (F-15C/D/E)	DLF	A	C
Engine, aircraft, F100-PW-229 (F15E, F16C/D)	EUF	A	C

Table 2.4—Continued

Marines			
Weapon System	**WSDC**	**2014 WSGC**	**2016 WSGC**
Computer system, DIG (TAMCN: A00337)	7AM	C	A
Mine-resistant vehicle (TAMCN: D00237)	17M	A	C
Cougar cat II surge	19M	A	C
Truck, wrecker, LVSR	1PM	A	C
Network management (TAMCN: A02437G)	4MM	A	C
LVSR cargo truck	8EM	A	C
Navy			
Weapon System	**WSDC**	**2014 WSGC**	**2016 WSGC**
Aircraft, Hawkeye E-2C	17N	A	C
Aircraft, E-6 Tacamo	20N	A	C
Helicopter, CH-53 D/E	41N	A	C
Engine, aircraft, J-52	49N	A	C
Engine, aircraft, T-64	50N	A	C
Littoral combat ship (LCS 2)	B2N	A	C
Littoral combat ship (LCS 1)	C1N	A	C

NOTE: The weapon system designator code (WSDC) is a three-character code uniquely identifying weapon systems participating in WSSP. The last character identifies the Service offering the system (A for Army, F for Air Force, N for Navy, and M for Marine Corps).

Volume of NIINs in the WSSP

One challenge in managing NIIN assignments to weapon systems—and, indeed, to achieving WSSP goals—is the very large number of NIINs that the services provide to DLA to help support designated weapon systems. Table 2.5 shows that the total NIIN population approaches 2.5 million. The table breaks out the population by NIIN priority, with the highest-priority NIIN being those with EC 1 (i.e.,

Table 2.5
WSSP NIIN Population, by Priority, 2016

National Stock Number Rank	WSIC	WSGC	EC	% of NIINs	Number of Demands in 2016
1	F	A	1	32.1	784,664
2	G	A	5	0.7	16,846
3	H	A	6	0.4	8,851
4	L	B	1	10.9	196,354
5	T	C	1	8.8	278,728
6	M	B	5	0.5	265,846
7	W	C	5	0.6	13,196
8	P	B	6	0.1	2,363
9	X	C	6	0.1	112,073
10	J	A	7	8.0	130,060
11	R	B	7	4.6	214,608
12	Y	C	7	5.1	14,882
13	K	A	3/blank	11.4	2,051
14	S	B	3/blank	5.3	123,948
15	Z	C	3/blank	11.4	279,568

SOURCES: Headquarters, U.S. Department of the Army (2009); and DLA Office of Operations Research and Resource Analysis (DORRA) (for current WSSP NIINs).

NOTE: WSIC is defined as the highest rank listing in the entire WSSP population for a given NIIN.

deadlines weapon systems) for WSGC A weapon systems, and so forth. Almost one-third of the NIIN population is at the highest level of priority. One-half of the NIIN population is in the top five tiers of priority (out of 15 tiers).

The Preponderance of High-Priority NIINs in the WSSP

The dominance of high-priority NIINs is even starker when levels of demand are taken into account. Table 2.6 shows the same ranking of WSSP NIINs as is shown in Table 2.5, but with frequency of demand displayed for 2016. The highest-priority NIINs account for 60 percent of all DLA issues. The top five tiers comprise over two-thirds of everything DLA manages and are issued from DLA depots. Indeed, only 10 percent of DLA issues of stocked items are for NIINs not part of the WSSP.

Table 2.6
DLA Issues by NIIN Priority, All DLA-Direct Issues, 2016

WSSP Priority	WSIC	EC	# of Issues	% of Issues	Cumulative %
1	F	1	4,037,681	59.8	59.8
2	G	5	101,765	1.5	61.3
3	H	6	48,820	0.7	62.0
4	L	1	235,896	3.5	65.5
5	T	1	212,589	3.2	68.6
6	M	5	15,121	0.2	68.9
7	W	5	20,989	0.3	69.2
8	P	6	4,595	0.1	69.2
9	X	6	5,691	0.1	69.3
10	J	7	323,923	4.8	74.1
11	R	7	57,026	0.8	75.0
12	Y	7	57,168	0.9	75.8
13	K	3/blank	693,166	10.3	86.1
14	S	3/blank	164,326	2.4	88.5
15	Z	3/blank	96,337	1.4	89.9
DLA issues not in the WSSP			681,156	10.1	100.0

SOURCES: DLA Strategic Distribution Database (SDDB); and DORRA (for WSSP NIINs).

A Breakout of WSSP NIINs by Service, Priority, and Demand Value

The WSSP NIIN population is not only skewed to the highest-priority items but also shows disproportionate demands among the services. Table 2.7 lists how demand for all DLA issues in 2014–2015 broke out by WSSP priority (or no priority, if the NIINs were not part of the WSSP). Over 1.8 million WSSP NIINs had no demands in that two-year period. The vast majority of DLA issues were for WSSP NIINs; only 14 percent of the value issued was for non-WSSP items. And two-thirds of the total demand value was for the highest-priority WSSP NIINs.

The assignment of WSSP NIINs as being of high priority, and the proportion of demands they account for, is heavily skewed by the services. Table 2.8 shows the number of NIINs assigned by each service into the five highest-priority WSSP categories, the number and value of DLA issues in 2014–2015, and the percent of all DLA issues they account for.[6] The Navy dominates assignment of high priorities, both by number of NIINs and the value of DLA issues. The Air Force also generates large numbers of high-priority NIINs, with the Army

Table 2.7
Value of WSSP NIIN Demand, 2014–2015, by Priority

Group	# NIINs	# Demands	Value ($billions)	% of total demand
Priorities 1–5	430,042	9,272,297	$7.399	66.3
Priorities 6–10	56,849	524,988	$0.650	5.8
Priorities 11–15	172,129	2,334,838	$1.535	13.8
Not in the WSSP	49,208	709,542	$1.569	14.1
TOTAL	708,228	12,841,665	$11.154	100.0

SOURCE: DLA Distribution System Support management information system issues data.

[6] The same NIIN can be assigned to priorities 1–5 by multiple services, yielding a higher total of NIINs and higher aggregate demand value than is shown in Table 2.6.

Table 2.8
Value of Demand for WSSP High-Priority NIIN, by Service Designating Priority, 2014–2015

Service	# High-Priority WSSP NIINs	Value of DLA Issues	% of all DLA Issues
Navy	263,020	$4.2B	38
Air Force	164,394	$1.6B	14
Army	77,951	$0.8B	7
Marine Corps	58,253	$0.8B	7

SOURCE: DLA Distribution System Support management information system issues data.

and Marine Corps assigning far fewer NIINs, with much less demand value, to this highest category.[7]

Since the intent of the WSSP is to direct investments and management attention to the highest-priority items driving weapon system readiness, these results suggest a potential skewing of resources away from the Army and the Marine Corps and toward the Air Force and Navy, and especially the Navy. As it turns out, this imbalance is not a significant concern, but arises only because the WSSP is not achieving its major purpose.

Subsequent Lack of Prioritization of Investment and Sameness of MA

As is commonly noted, when *everything* is a priority, *nothing* is a priority. With the large majority of WSSP demands being designated as of a high priority, we would anticipate DLA having difficulty making investment decisions to effectively prioritize the MA of high-priority NIINs. And in fact, historical performance confirms this. Figure 2.1 shows MA by WSSP priority ranking for 2016 issues from DLA depots. Investment allocation by WSSP priority would yield a downward-

[7] In the same time period, the Army accounted for 38 percent of all DLA issues, and the Marine Corps for 12 percent. The Navy accounted for 31 percent, and the Air Force for 19 percent.

Figure 2.1
MA, All DLA-Direct Issues, 2016

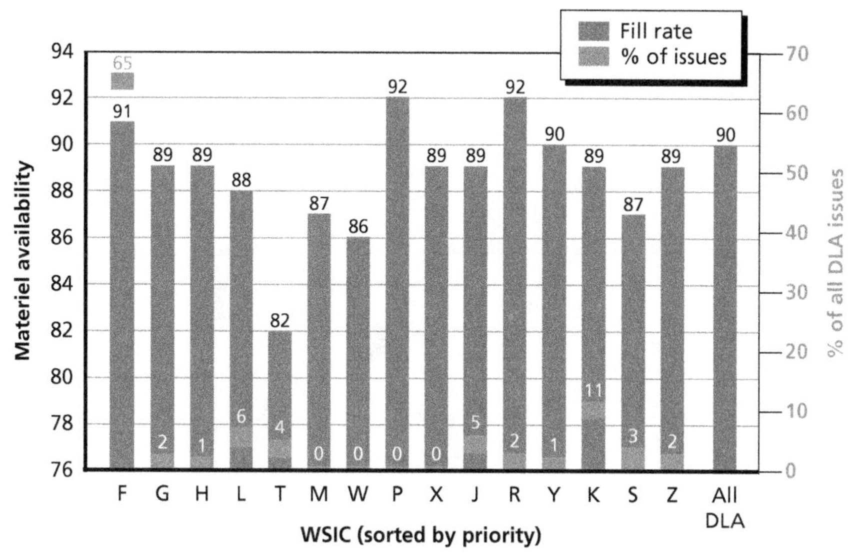

SOURCES: DORRA (for WSSP NIINs); and SDDB.
NOTE: Excludes DLA-executed issues to Air Logistics Complexes.
RAND *RR2496OSD-2.1*

sloping trend. Instead, performance is randomly distributed around the 90-percent level—the aggregate DLA performance.[8] In essence, then, the WSSP is not able to achieve its major aim of producing higher availability for the most critical items. As will be shown later in Figures 4.2 and 4.3, DLA bases MA not on the assigned priority of the part, as done through the WSSP, but more typically by the cost of the item.

---

[8] MA for the 10 percent of demand for non-WSSP items is much lower (at 80 percent), as these NIINs tend to be low- and variable-demand items, where it is difficult to buy enough stock to hit higher availability levels.

CHAPTER THREE

An Alternative WSSP Approach

An Overview of the RAND Approach

The largest obstacle to the success of the WSSP is an excessive number of NIINs, especially high-priority ones, in the system that are of questionable relevance to weapon system readiness, and an investment strategy that seeks to achieve overall MA goals with little consideration of how to direct funds to the most critical items. An alternative approach for making the WSSP effective would do two things: first, it would present a method for determining "true" critical RD NIINs, with the expectation that these critical NIINs form a much smaller part of the overall population; and second, it would provide rules to direct proportionately more investment toward the critical RDs, even at the expense of less critical items. In this chapter we offer a way to identify critical RDs based on U.S. Army and Air Force experience.[1]

True Readiness Drivers

As previously noted, WSSP priorities are based on the EC, typically determined early in the weapon's life cycle and during initial provisioning—well before any experience in the field. These assignments may

[1] Our research is intended as a proof of concept, using Army and Air Force data to illustrate an alternative approach. Given RAND's long history of Army and Air Force logistics research and deep familiarity with the data analyzed in this project, it was decided that focusing on these two services would provide the most effective approach to analyzing alternative approaches for the WSSP.

be right or wrong in identifying parts whose failure would deadline weapon systems, but absent actual field data it is hard to judge the relevance of the assigned EC.

The actual impact of parts on weapon system readiness, we argue, is best determined from that field experience—from the deadlining reports field mechanics maintain to indicate what broken parts need to be replaced before a system can be pronounced as fully ready for assigned operations. The method we used to assign NIIN criticality draw from maintenance data for the Army and the Air Force. The RAND team used three different data sources:

1. The Army's Equipment Downtime Analyzer (EDA): The EDA captures reportable systems deadlining parts reports submitted by field maintenance personnel and creates histories of deadlined systems repairs. It is limited to nonaviation systems (Peltz et al., 2002).
2. The Army Aviation Aircraft on Ground (AOG) database: The Army Aviation and Missile Command's AOG system captures and archives requisitions for deadlining parts for Army Aviation; AOG requisitions are based on direct call-ins and bypass standard requisitioning channels in an effort to highlight their priority and achieve the fastest response times (Headquarters, U.S. Department of the Army, 2010, pp. 5–6).
3. The Air Force Logistics, Installations and Mission Support–Enterprise View (LIMS–EV) database: LIMS–EV is a business intelligence environment that integrates logistics and installation information from across the Air Force enterprise. The system contains data on weapons system availability, munitions, vehicles, support equipment, and supply chain management status (Petcoff, 2010).

Each of these data sources stores transactional maintenance data, but both the data formats and time windows from which they are drawn are different. The EDA database tracks maintenance history and identifies part requisitions resulting in a *not mission capable* end item status. It then aggregates these data and creates a NIIN-to-NIIN matching between all parts and end items. After that, each part NIIN is designated as a critical or noncritical RD based on the number of not mission capable requisitions it has against a given end item. The EDA is based on three years of maintenance history because this is what is considered relevant for deciding on-hand quantities for authorized stockage lists at Army supply support activities.

The Air Force tracks all parts that are required to keep a system mission capable. When a part is available through the local supply system (i.e., base stock), the system is reported as *not mission capable, maintenance* while the part is being repaired or replaced. When a part is not available through the local supply system, the system is reported as *not mission capable, supply*, a requisition is created to order the part, and a Mission Impaired Capability Awaiting Parts (MICAP) incident is recorded. There is also a *not mission capable, both* code, which should be used if there are *not mission capable, supply* parts on order and maintenance is still ongoing (typically used when an aircraft is in phase inspection).

MICAPs and associated supply codes are how the Air Force signals to the supply system which parts should be prioritized and how fast the system needs to respond. There are several ways a MICAP can be satisfied, including through lateral support from another base's stock, through cannibalization, or through DLA retail supply. Since this analysis aims to help DLA better prioritize across NIINs for retail supply, we focus only on MICAP data and not the broader set of parts required to keep a system mission capable, since these are the parts for which the Air Force has not maintained sufficient levels of base stock and, by nature of a MICAP designation, the Air Force is signaling that the part should be prioritized.[2]

---

[2] Since MICAP actions are based partly on local stock availability and not on some more "fundamental" relationship of the part to the airframe (as would be the case if the MICAP were initiated by the failure of the part and not its lack of local availability), it may be that our using MICAP data reflects those with temporary supply problems, and diverting funds to them might simply cause "new" MICAP NIINs to emerge. We believe this is not the case, as the population of MICAP NIINs tends to stay stable over time, suggesting the problem is beyond any temporary lack of supply. See the section below, "The Stability of RD NIINs."

Critical RDs are NIINs that are especially crucial in driving weapon system readiness and imposing challenges on the supply system. These drivers dominate deadlining actions; our aim is to identify the smallest population of NIINs that account for the largest number of deadlining actions.[3] We identify these critical RDs based on a set of criteria and thresholds similar to the one the Army uses to establish brigade-level stocking policies (Girardini, Lackey, and Peltz, 2007); consider part X and weapon system Y:

- A part must have at least five deadlining demands against a specific weapon system in the last three years.
- That part must account for at least 0.5 percent of all deadlining demands for the weapon system:

$$[(\text{\# deadlining demands for part X against weapon system Y}) / (\text{total \# deadlining demands for weapon system Y})] \times 100 > 0.5 \text{ percent.}$$

- At least 3 percent of all demands for part X have to be deadlining:

$$[(\text{\# of deadlining demands for part X across all weapon systems}) / (\text{total demands for part X by the service, deadlining or not})] \times 100 > 3 \text{ percent.}$$

EDA data are only used for ground systems because aviation units have not yet transitioned to the Army's current supply chain enterprise resource planning system known as Global Combat Support System Army. For Army Aviation, the RAND team used the AOG data set, which comprises all records of broken parts leading to an aviation asset being kept "on the ground." The RAND team was able to acquire a ten-year archive of all AOG actions, from 2005 through 2015 during Operation Enduring Freedom and Operation Iraqi Freedom.

For the Air Force, the RAND team used MICAP data, which were pulled from LIMS–EV over a period of five years (2012–2016).

---

[3] We generally set the target as smallest number of NIINs accounting for at least 50 percent of deadlining parts demands.

This duration was chosen because it represented a similar volume of data as the three years of EDA data and ten years of AOG data.

While the EDA's threshold criteria were defined years ago, the MICAP and AOG did not come with a built-in RD threshold. We had to define the threshold in order to define Army Aviation and Air Force RD NIINs. Our aim was to create criterion that mirror the EDA logic, but the data that the research team possessed only allowed us to accurately mimic threshold 1 from the EDA list above. To do so, we used the following criteria:

- For the AOG, parts with more than ten deadlining events against a given aircraft over the ten-year period were considered to be critical RDs.
- The threshold and criteria we set for the MICAP data were based on the Urgency Justification Code (UJC). A MICAP with a UJC of 1M or 1A means that the MICAP resulted in the weapon system/engine being *not mission capable, supply*. A UJC of JA means that the weapon system/engine was partially mission capable because of the MICAP. For a NIIN to be considered a critical RD for an Air Force weapon system there must have been at least ten UJC 1A/M or JA MICAPS over the course of the five years (2012–2016).[4]

How Well Do RD NIINs Line Up with the EC? The Case of the M1 Abrams Tank

When comparing the population of EC 1 NIINs and RD NIINs for the Army, there were a few glaring differences. Our RD list was much shorter, and all our NIINs had deadlining occurrences associated with them by the very nature of our method. The EC 1 NIINs population included many NIINs that were not RDs and did not see a deadlining demand in the three years of data that we collected. On the other hand, there are also many RDs that are not EC 1 NIINs. This is

[4] We further set the criteria for critical MICAP NIINs so that no more than 25 percent of these MICAPs were UJC JA, or causing a partially mission-capable status. This was to ensure that while some weight was given to NIINs causing partial mission capability, most of the MICAPs result in true *not mission capable* status.

common across most Army WSGC A systems, but to help illustrate the point, here is a closer look at the Abrams Tank.

As shown in Figure 3.1, of the 3,370 EC 1 Abrams NIINs, 2,865 never saw a deadlining demand in 2016, 400 got at least one deadlining demand but were not considered critical RDs, and 106 were critical RDs.

Figure 3.2 shows the breakout of the Abrams tank's 430 RD NIINs. Of the 430 RDs, 106 were assigned the highest WSSP rank (see Table 2.5 for the rankings), 20 had the second rank, 43 of our critical RDs were assigned rank 10 in the WSSP, and 100 had rank 13 (of 15). There were 160 critical RDs for the tank that were not in the WSSP at all.

The Stability of RD NIINs

ECs, which are generally set during acquisition, are by definition fairly stable. Can the same be said for NIINs that have been tagged as critical RDs based on field maintenance data? The evidence from Table 3.1 strongly suggests both that critical RD NIINs are stable year by year

Figure 3.1
Most M1 Abrams Tank WSSP EC 1 NIINs Are Not RDs

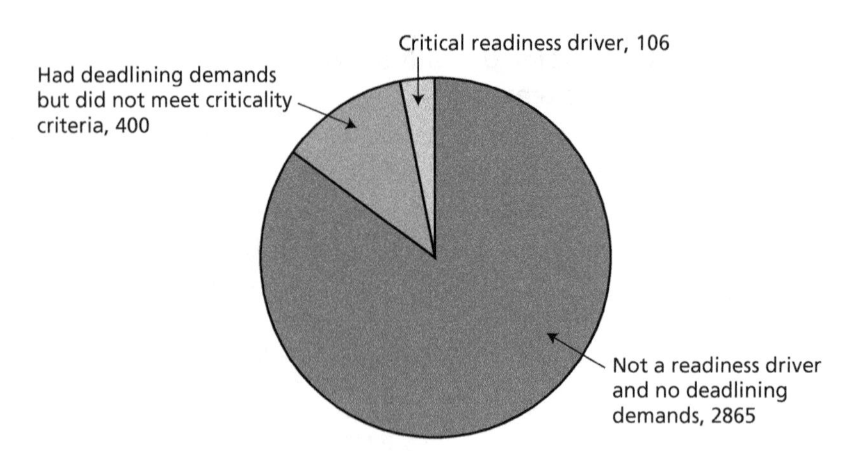

SOURCE: DORRA (for WSSP NIINs); and EDA.
RAND *RR2496OSD-3.1*

Figure 3.2
One-Fourth of Abrams Tank RD NIINs Have High WSSP Priority

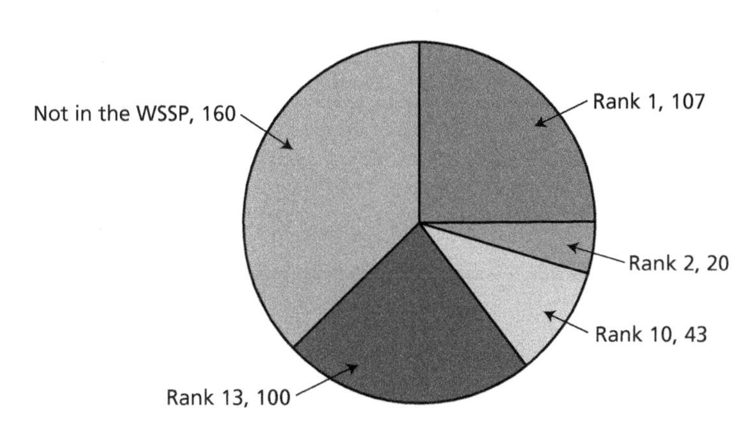

SOURCE: DORRA (for WSSP NIINs); and EDA.
RAND *RR2496OSD-3.2*

and that a small number of those NIINs account for a large percentage of system deadlining events.

The table is based on all Air Force MICAP demands over six years (2011–2016). It defines a high-volume NIIN as one in the smallest population of NIINs that accounts for one-half of all MICAP actions in a given year. The table shows how many years NIINs continued to be in the top half of the population of MICAP actions, with a maximum of six years.

Almost 87 percent of NIINs never appear in the top half of the RDs in any of the six years, accounting for about one-half of MICAP actions. However, a very small percentage of NIINs appear year after year—some being demand drivers for each of the six years. Less than 3 percent of the NIINs account for almost 30 percent of all MICAP actions. A small handful—just 277 NIINs, or 0.5 percent—were a top driver for all six years and accounted for more than 11 percent of all MICAP demands.

In short, a small population of NIINs account for almost one-half of the Air Force MICAP population and generally tend to cause MICAP year after year.

Table 3.1
Air Force MICAP NIIN Actions, 2011–2016

Years with NIIN in the Top Half of MICAP Actions	# NIINs	Total MICAP Actions	% of All NIINs	% of All MICAP Actions
NIINs never in top half	51,327	187,614	86.8%	51.5
1	4,884	38,689	8.3%	10.6
2	1,449	30,791	2.4%	8.5
3	706	27,520	1.2%	7.6
4	338	17,490	0.6%	4.8
5	258	21,258	0.4%	5.8
6	277	40,797	0.5%	11.2
Total	59,239	364,159		

SOURCE: LIMS–EV.

Reducing the WSSP NIIN Population by Using Field Maintenance–Based Critical RDs

Adopting a field maintenance–based approach for identifying critical RD NIINs for participation in the WSSP would yield two major benefits:

1. It would more directly tie WSSP priorities to weapon system readiness by limiting the WSSP population to NIINs that have proven relevance to maintaining readiness.
2. It would allow for a much smaller, and more manageable, population of critical NIINs.

Such an approach would provide an empirical basis for identifying critical items that existing means do not deliver. Engineer estimates made during product development or during initial provisioning prior to field experience, while no doubt made in good faith, will not reflect actual experience; further, because those estimates are not tied to actual field events, there is little pressure to limit the number of NIINs considered critical. As a result, we see an explosion of high-priority NIINs both in the WSSP and in general.

Based on field experience, far fewer NIINs are actually critical RDs. Using that field experience, the services can ensure that the minimum number of NIINs having the maximum impact will be so considered. In the approach we have adopted here, we have identified the smallest number of NIINs that account for at least one-half of all deadlining events in the period studied. That is, among all NIINs that contribute at least some to readiness, we have identified the most critical ones.

This limitation is vital. The list of RD NIINs could be increased enormously by loosening these criteria. For example, trying to cover, say, 75 percent of deadlining events would double the number of NIINs, most of which would contribute very few deadlining actions. Giving these NIINs priority in replenishment investments would yield a very poor readiness return on the money spent: there are far too many NIINs to cover, the vast majority of their demands are not readiness-related, and they only infrequently cause the deadlining of a system. Part of our recommendations (in Chapter Seven) are to establish a mutually agreed-upon set of criteria for limiting the number of NIINs to the most critical RDs to thus prevent imbalances in those NIIN sets—and therefore funds to be invested—across the services.

Tables 3.2 and 3.3 show how much smaller the critical RD population is than the current high-priority NIIN populations in the WSSP. Table 3.2 shows that for important Air Force systems, using our MICAP event–based approach for identifying critical RD, NIINs would shrink the NIIN population by 99 percent. Table 3.3 shows the situation for the Army, where the list would shrink by almost 90 percent.

Having smaller, more targeted populations would not only make investment strategies more feasible and productive by tying investment dollars directly to empirically based readiness events but also might reduce management workload by greatly reducing the number of NIINs to deal with.

Table 3.2
Comparison of Air Force High-Priority WSSP NIINs and MICAP-Based Critical RDs

WSGC	Weapon System	# of WSSP High-Priority NIINs	# of Critical MICAP NIINs
B	Aircraft, F-16	38,746	1,669
B	Aircraft, Eagle F-15	59,931	871
A	Aircraft, Stratolifter C/KC-135	48,039	690
A	Aircraft, Hercules C-130	34,535	507
B	Aircraft, Thunderbolt II, A-10	27,468	335
A	Aircraft, Osprey CV-22B	7,682	160
B	Aircraft, B-1B	26,625	138
B	Helicopter, HH-60 all variations Pave Hawk	2,149	120
A	Aircraft, Stratofortress B-52	35,403	111
A	Aircraft, Airlifter C-17A	8,374	94
A	Aircraft, Galaxy C-5	46,372	72
A	Aircraft, SOF (C-130H, AC-130J, AC-130U, EC-130E, EC-130H, HC)	46,411	43
A	F-22 Raptor air dominance fighter	1,930	12
A	Aircraft, B-2 bomber (ATB)	13,370	5
A	Global Hawk high-altitude-long endurance unmanned aerial vehicle	1,658	4
A	Missile, Minuteman III, LGM-30	27,346	1
A	U-2 Airframe	3,591	1
Total		429,630	4,833

SOURCES: DORRA (for WSSP NIINs); and LIMS–EV.

NOTE: WSIC F and L (EC 1 for WSGC A and B systems, respectively).

Table 3.3
Comparison of Army High-Priority WSSP NIINs and MICAP-Based "True" RDs

WSGC	Weapon System	# of WSSP High-Priority NIINs	# of Critical RD NIINs
A	AH-64E	12,250	2,236
A	AH-64-D Longbow	4,383	632
A	Helicopter observation, OH-58D	4,394	246
A	Helicopter, Black Hawk, UH-60A	8,988	1,863
A	Helicopter, Chinook, CH-47	7,136	974
A	Family of medium and light tactical vehicles	4,061	954
A	Fighting vehicle systems, Bradley	3,883	174
A	Howitzer, M-109 series	1,941	125
A	Infantry carrier (Stryker fighting vehicle)	186	68
A	Tank, Abrams M-1	4,000	241
A	Parent WSDC for all Army Mine-Resistant, Ambush-Protected WSDCS	8,172	784
A	Truck, heavy expanded mobility tactical	1,730	693
A	Truck, vehicle system, 1 1/4-ton (HMMWV)	3,587	788
Total		78,622	9,778

SOURCES: DORRA (for WSSP NIINs); EDA; and Armament, Munitions and Chemical Command (AOG) data.

NOTE: Highest-priority EC NIINs only (EC 1, 5, 6); total WSSP NIINs for these systems = 231,034.

CHAPTER FOUR

Modeling Alternative Approaches: An Army/Air Force Case Study

Overview

This chapter presents an approach for assessing the potential benefit of a critical RD strategy to prioritize DLA's stock investments and thereby provide higher weapon system support. The key task is identifying true critical RDs, as presented in Chapter Three. Once identified, those NIINs will be given preference in stock investment decisions aimed at increasing safety stock levels (i.e., the amount of additional stock needed to cover unexpected surges in demand and to yield a targeted MA). Given a fixed cost restraint, fewer funds would be invested in noncritical RDs on a one-time basis—that is, their safety stock levels would be allowed to decline and a lower MA would be accepted for these lower-priority parts.

Our proof of principle case is limited to Army and Air Force critical RDs, with the aim of demonstrating the validity of the approach; a complete analysis would include critical RDs from all four services. Lacking equivalent lists from the Marine Corps and the Navy, we configured the analysis to test the concept under fair conditions. That means that the potential funds for reallocation—that is, funds connected with NRD NIINs—have to be limited to DLA items that go exclusively (or almost exclusively) to the Army and the Air Force. While in principle this approach could be implemented as we depict it here for those two services, the approach will work best once all four services are able to generate critical RD lists based on approaches that are mutually agreeable across the enterprise.

Data

Data sources and the methodology for determining the list of critical RD NIINs were discussed in Chapter Three. All critical RD NIINs for Air Force weapon systems with WSGC A and B, and all Army systems with WSGC A, were included.[1] Demand data come from DLA's Distribution System Support management information system, an archived record of all issues from DLA depots. Our analysis was limited to all items requisitioned from DLA depots for the years 2014–2015. Unit prices for all NIINs were from the Federal Logistics Data on Mobile Media (also known as FEDLOG).

General Statistics on the Demand Population

Table 4.1 shows the overall demand level and number of NIINs for Army/Air Force critical RDs and NRDs in 2014–2015, and the distribution of their unit prices. Note in particular that Air Force RDs tended to be more expensive, while Army RDs had far more demands.

Table 4.2 shows the aggregate values for demands, quantity demanded, and value of the demands in the test period. RDs accounted

Table 4.1
Number of and Demand for Army/Air Force RDs and NRDs, and Distribution of Unit Prices, 2014–2015

	# NIINs	# Demands	Mean	95th Percentile	75th Percentile	Median
Army RDs	9,494	1,996,475	$182	$748	$82	$13
Air Force RDs	4,956	363,048	$1,536	$6,500	$1,307	$290
NRDs	190,312	3,889,693	$342	$1,341	$152	$36

SOURCES: AOG data; DORRA; EDA; LIMS–EV; AOG, and SDDB.

[1] Air Force WSGC B systems were included because they feature important aircraft (such as the F-15 and F-16), while many of the Air Force WSGC A systems have few or no MICAP demands (e.g., several ground support systems, such as for the C-130) or do not yet demand much organic DLA materiel (e.g., the F-22).

Table 4.2
Aggregate Demand for Army/Air Force RDs and NRDs, 2014–2015

	# Demands	Quantity	Value	Demands %	Quantity %	Value %
Total	6,249,216	107,314,599	$5,599,188,713			
NRDs	3,889,693	63,029,749	$3,674,892,774	62	59	66
RDs	2,359,523	44,284,850	$1,924,295,939	38	41	34

SOURCES: AOG data; DORRA; EDA; LIMS–EV; and SDDB.

for just over one-third of the total value of Army/Air Force demand of DLA materiel.

Total demand value is dominated by more expensive NIINs, as illustrated in Figure 4.1.

Figure 4.1
Two-Year Demand Value for Critical and Noncritical NIINs, by Unit Price Group

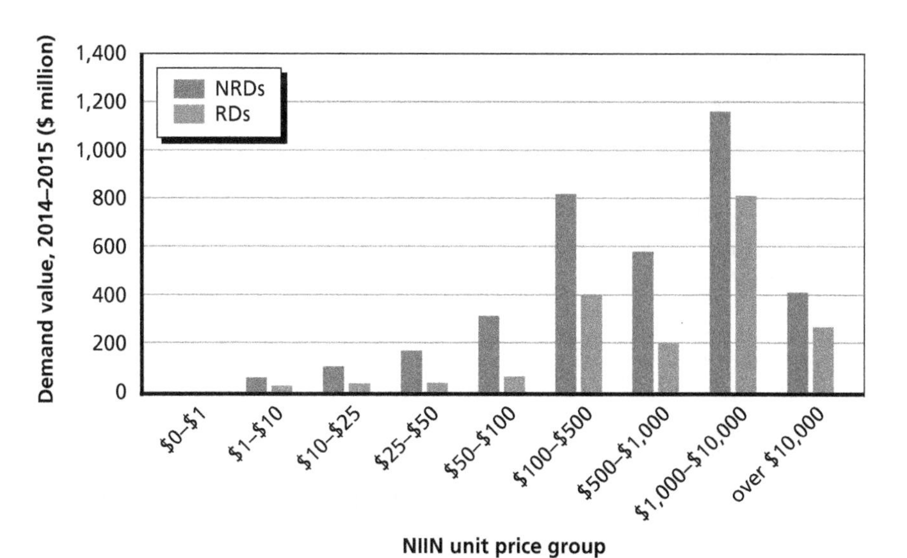

SOURCE: AOG data; DORRA; EDA; LIMS–EV; and SDDB.
RAND *RR2496OSD-4.1*

The Simulation Approach

The simulation is designed to be a proof of concept regarding the trade-off between RD and NRD investment allocation and MA. The simulation does not strive to replicate DLA's procurement logic, nor does it intend to forecast MA as a function of actual investment allocation. The simulation's purpose is to explore the potential for MA versus investment allocation trade-offs as a precursor to exploring real-world investment reallocation and MA effects.

Though we are not privy to DLA's complete and highly complex replenishment logic, we quantify the empirical MA performance that the logic yields. In general, MA is very high for inexpensive NIINs and much lower for expensive NIINs, resulting in an overall dollar unweighted MA of approximately 90 percent. The simulation is calibrated to mimic observed MA by unit price group and then used to model hypothetical investment scenarios.

There are three steps used to build and use the simulation for this purpose:

1. Create a computer simulation that models inventory, replenishment to DLA, and issues from DLA using observed DLA transaction data.
2. Calibrate the procurement logic in the simulation model to have the same MA results by unit price group as displayed in Figure 4.2.
3. Simulate hypothetical changes in stock investment where increased investment in RD NIIN stock is decremented from the NRD NIIN available procurement funds.

Once calibrated (i.e., step 2), we run hypothetical scenarios where available funds for NRD replenishment are reallocated and used to increase safety stock of RDs (step 3). This has the effect of increasing MA for RDs and decreasing MA for NRDs. We express the reallocation value as a percentage of the annual RD demand. Specifically, the following are the hypothetical scenarios that we simulate in which the dollar value relocated from NRD procurement funds to additional RD safety stock is:

A. 20 percent of the annual RD demand value
B. 40 percent of the annual RD demand value.

For example, if RD annual demand value is $500 million, then $100 million is removed from the NRD Obligation Authority (OA), and $100 million is used for additional RD safety stock in scenario B ($500 million × 20 percent = $100 million).

Simulation Mechanics and Assumptions

The simulation is a stock-and-flow model that reflects replenishment logic, replenishment receipts, demands and issues, and stock levels. It is designed to replicate system performance to be able to assess how that performance would change under different scenarios. The simulation is designed to show the directional effect on RD and NRD MA given shifting procurement funds. It is not designed to replicate DLA procurement logic and predict the exact effect of such an approach. Model details and assumptions are explained in the Appendix.

Step 1: Model Creation and Assumptions

The simulation models the inflows, outflows, and stock levels by NIIN and calculates performance metrics. When the stock levels are reviewed in the model, the procurement logic decides how much to procure from vendors and submits a replenishment request. Given a Production and Administrative Lead Time (PALT), the replenishment is observed and the stock is increased by the replenishment quantity. If stock is available and a demand event occurs in the simulation, the simulation will issue stock to customers thus decreasing stock on-hand. As the simulation time advances, the model keeps track of performance metrics including the number of filled demands, due out quantities, demand quantities, and replenishment request quantities. These metrics allow post processing that calculates summary statistics about model performance including MA by NIIN and unit price group.

Assumptions are necessary and are made regarding procurement logic and replenishment. We do not have visibility into DLA's procurement decisionmaking and logic and therefore must assume a reasonable replenishment logic to perform this proof of concept.[2] We assume an order-up-to policy with periodic review, a common ordering policy in practice and in inventory theory. In this policy, the desired replenishment quantity is the difference between a Requisition Objective (RO) target and the current inventory position (IP), where the IP is the sum of the quantity on hand and the difference of the due-in and due-out quantities (i.e., replenishment quantity = RO – [on hand + (due in – due out)]). We assume that the replenishment request quantity is decided each week in the simulation.[3] The desired replenishment value is the dollar value associated with the desired replenishment quantity.

In the simulation we make assumptions so that the model performance is that of a steady-state system rather than the transient short-term effects of the current system if investments are reallocated. The goal is to demonstrate the proof of concept in the steady state and not as a function of current on-hand or vendor capabilities. The transient effects that are intentionally excluded are the effects on MA due to current on-hand quantity, current vendor on-hand quantity, production capabilities, and back order status. Therefore, we start the simulation on-hand quantity equal to the RO to remove the effect the current on-hand quantity has on near-term performance. As simulation-time progresses, the on-hand quantity decreases below the RO and eventually has no trend. Once the trend is not observable, then we start to collect performance statistics to evaluate performance in the steady state.[4] We also assume a 26-week PALT for all NIINs, which is approximately the average PALT across all NIINs actually experienced.[5] Depending on

---

[2] If this effort were intended to be more than a proof of concept where actual performance changes were predicted based on procurement logic changes, then a thorough modeling of DLA's logic would be necessary.

[3] In actual practice, DLA considers replenishment decisions in real time depending on the OA, on-hand quantities, and priorities, among other attributes.

[4] This technique is called removing the initial bias of a simulation and is common practice in practical and academic applications of simulation where steady-state analysis is desired.

[5] PALT ranges widely across the NIIN population, based on contract type (e.g., long-term contract or spot buy) and the challenges of production. For detail on the range of PALTs, see Peltz et al. (2015, Appendix G).

the vendor on-hand quantity, production capabilities, and back order status, the PALT can and does change over time. Given that the vendors' effect is beyond the scope of this research and our desire is to not confound the vendors' effect on MA with the investment effect, in this proof of concept we thus assume a universal 26-week PALT. We then simulate the first 52 weeks and discard the performance so the simulated on-hand quantity can reach a steady state before performance statistics are captured; this removes the initial bias due to starting the on-hand quantity at the RO.

Step 2: Calibrating the Model

Before running hypothetical investment scenarios, it is necessary to calibrate the model so that its performance reflects reality. As mentioned, we do not model DLA's complete procurement logic because of its complexity and our lack of access to the voluminous rules employed. However, we can and do calibrate the model such that the assumed order-up-to policy has a simulated MA performance that mimics reality by unit price group. To calibrate, we adjust the ROs for each unit price group such that the simulated MA for each is approximately equivalent to the actual observed MA. This calibration is performed iteratively where simulated MAs are calculated, and ROs are adjusted repeatedly until the simulated MA is similar to that of the observed MA.

It is important to calibrate such that the starting point for the hypothetical scenarios has at least similar performance to that of the real world. The hypothetical scenarios are naturally an extrapolation from the calibration but are defined to progressively extrapolate from the calibration to simulate the MA performance.

Figures 4.2 and 4.3 indicate how closely the calibration matches observed MA. Overall, the model matches MA performance in 2014–2015 perfectly, with 90 percent MA in both cases. Figures 4.2 and 4.3 break down actual and calibrated modeled results in more detail, showing MA in terms of whether the NIIN was in the RD or the general population and by NIIN unit price groups. Figure 4.2 shows actual MA in 2014–2015 broken out by NIIN population and nine groupings by increasing unit price. The figure shows a "staircase" pattern by which MA gradually declines as unit price increases, in line

Figure 4.2
Actual MA for DLA Issues, 2014–2015: RD and NRD NIINs

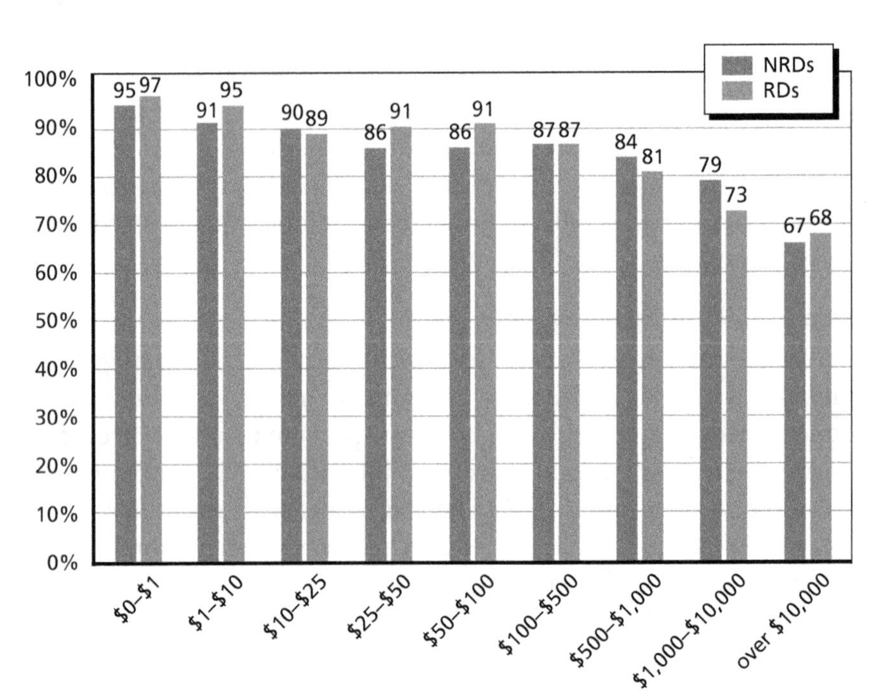

SOURCE: AOG data; DORRA; EDA; LIMS–EV; and SDDB.
RAND *RR2496OSD-4.2*

with efficient inventory investment strategies in pursuit of an aggregate goal (such as an overall MA rate of 90 percent). It also shows no significant difference in MA between RDs and NRDs, as can be expected since DLA has not targeted the RD NIINs for increased investment.[6,7]

---

6 Less expensive RDs do tend to have a somewhat higher MA than NRDs. This may derive in part from these NIINs having generally higher and more stable demand patterns (lessening the chance of demand spikes) and from a DLA policy of buying and stocking some level of inventory for high-priority WSSP NIINs assigned Acquisition Advice Code, Z ("nonstocked"), an action typically not taken for lower-priority items.

7 In order to find a successful calibration within a reasonable time, we stopped searching for a better calibration if the simulated MA within each unit price group is within ±1 percentage point. Since this model is not being used to predict real-world performance and is

Figure 4.3
Simulated MA for DLA Issues, 2014–2015: RD and NRD NIINs

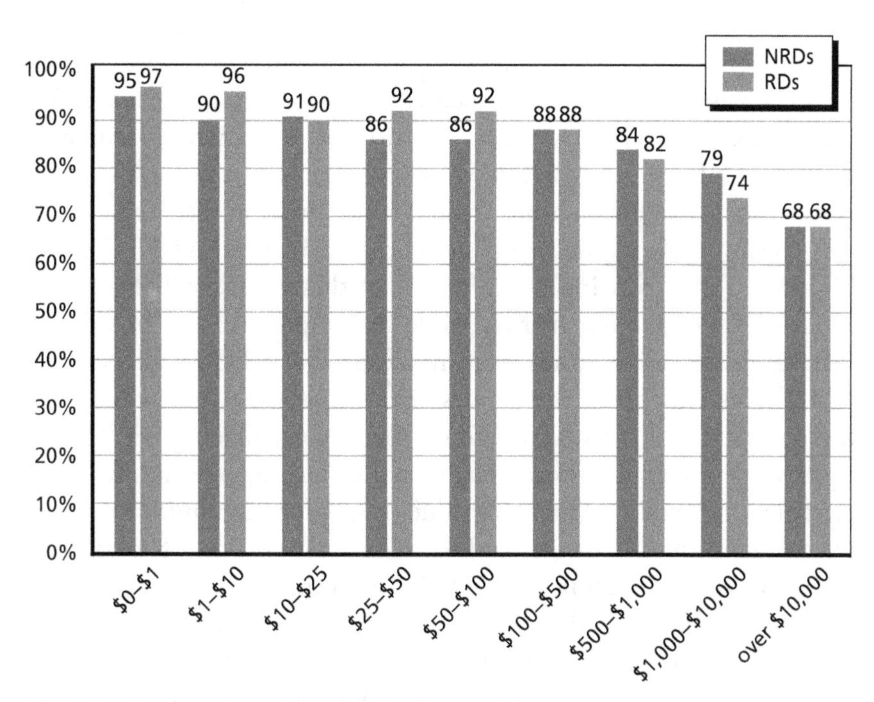

SOURCE: AOG data; DORRA; EDA; LIMS–EV; and SDDB.
RAND *RR2496OSD-4.3*

Figure 4.3 shows simulation results with the same unit price groups displayed in Figure 4.2.

In general, this model closely reflects real-world results, adding confidence to the potential real-world application of the alternative scenarios we will discuss shortly.

Step 3: Modeling Hypothetical Investment Scenarios

In the hypothetical scenarios where investments are diverted from NRD available procurement funds to RD safety stock, we implement a logic that decrements the former by the latter's value. For this purpose,

---

designed to demonstrate a proof of concept, this closely calibrated model is sufficient for this purpose even though it is not calibrated exactly to 2014–2015 performance.

we assume that the available procurement funds in the simulation for the NRD NIINs are the annual observed issue value for those NIINs; in other words, we assume that the available procurement funds are exactly equal to the issue value in the CY 2014–CY 2015 data so as to replace issued stock with procured stock.[8] To maintain high MA for less expensive NRD NIINs, we do not decrement procurement funds for NIINs below $100. We chose the $100 threshold—which is admittedly somewhat arbitrary and could be set at different, though still cheap, levels in future policy—because inexpensive NIINs have high MA, and reducing procurement of these items will likely have a significant reduction in MA while not improving the MA of RD NIINs. Reallocation funds from more expensive NRD NIINs to all RD NIINs will bring the greatest increase of RD MA while limiting the impact of NRD MA. To do this, we calculate the RD investment value for each hypothetical scenario (i.e., 20 percent or 40 percent of the annual RD demand value). Then for each unit price group we calculate the NRD demand value as a proportion of total NRD demand value, excluding those NRDs with unit price less than $100. That proportion is used to allocate the NRD procurement funds decrement. For example, in the Army and Air Force simulation 20 percent additional investment scenario, the total RD safety stock investment is $243 million. Additionally, the annual NRD procurement funds for unit price group of $25–$50 is $84 million and is 5 percent of the total NRD annual procurement value. Then 5 percent of the $243 million will be decremented from the $25–$50 NRD procurement funds for net procurement funds of $72 million (i.e., $84 million – 5 percent × $243 million; $84 million – $12 million = $72 million).

The decrement of NRD procurement funds does not leave enough funds to fully replace demand in the hypothetical scenarios. Therefore, we developed logic to allocate the available procurement funds to the NIINs that need the most replenishment value within each unit price band each simulated week. This construct has the important attribute

[8] Issue value should not be confused with inventory value. Inventory value is the value of the on-hand stock. Issue value is the value of stock that is issued from DLA to DLA customers.

that every NIIN within each category will continue to be procured throughout the simulation because if a NIIN is not procured for a few weeks, it will rise in priority as demands continue and the desired replenishment value continues to increase.[9]

For example, suppose desired procurement quantity is five for NIIN A, with a unit price of $10, and two for NIIN B, with a unit price of $20. Then the desired replenishment value for NIIN A is $50 (i.e., 5 × $10) and NIIN B is $40 (i.e., 2 × $20). Therefore, NIIN A has a higher procurement priority according to the logic. Now suppose that there is only enough available procurement funds to procure NIIN A for $50. In this case, NIIN A is procured and NIIN B is not procured in this simulated week. At this point, the desired replenishment value is $0 for NIIN A (it was procured in this simulated week) and $40 for NIIN B (it was not procured in this week). In the next week, we see no demand for NIIN A, and two more demands for NIIN B, increasing their desired procurement to $80 (i.e., $40 + 2 × $20). At this time, NIIN B has a higher priority over NIIN A, because NIIN B's desired replenishment value is $80 and NIIN A's desired replenishment value is $0. This approach prevents NIINs from continually being neglected each week because each week a NIIN is not procured it rises in procurement priority in the subsequent week.

---

[9] If a different prioritization logic is used, one that prioritizes procurements for NIINs with smaller unit prices, the NIINs with larger unit prices within each category will continually be neglected. This is not realistic given that DLA does not appear to continually neglect procurement for some NIINs.

CHAPTER FIVE

Results of the Army/Air Force Case Study

Reallocating OA from the Investment Base to RDs

In this chapter we seek to show the benefits of redirecting OA investments to RDs. We do so by restricting the "investment base" from which OA is redirected.

As noted in Chapter Four, we define that investment base as Army/Air Force–dominated NRD NIINs. By "Army/Air Force–dominated" we mean that at least 90 percent of demand in our test period (2014–2015) came from these two services, and by " NRD NIINs" we mean all DLA-direct NIINs (i.e., excluding direct vendor delivery items) that do not meet the threshold to be defined as critical RDs. As Table 4.2 showed, about two-thirds of demand value was in the NRD population. We simulated a single year, dividing that demand by two. Figure 5.1 illustrates the breakout of that $2.8 billion yearly demand by population group. Note that just under $100 million of the RD demand comes from the Navy and Marine Corps, and that amount may or may not be used for fixing their deadlined systems.[1] For the purposes of this simulation, some OA for Army/Air Force demand will go to support Navy/Marine Corps demand. In a comprehensive solution—were we able to identify Navy and Marine Corps critical RDs—the same process would work in reverse, resulting in cross-subsidies of service RDs. Until that comprehensive approach is feasible, and just for the purposes of this demonstration, we assume that the Army and Air

[1] For example, 10 percent of demand for the standard vehicle storage battery, the Hawker (a major RD for Army systems), is from the Marine Corps.

Figure 5.1
Yearly Demand Value for Air Force/Army-Dominated NIINs and Critical Readiness Drivers

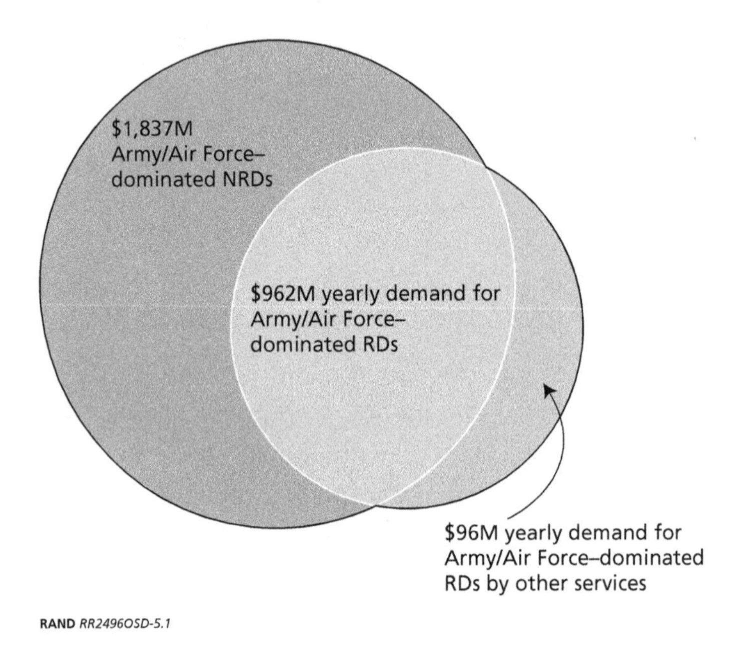

RAND *RR2496OSD-5.1*

Force would accept subsidizing demand from the other two services if the benefit to them was higher MA for their most critical needs.

Alternative Investment Strategies

We examine three cases for investing more in critical RDs:

- The base case: the current investment approach, with no preference given to critical RDs.
- A one-time increase in investment in critical RD safety stock, equal to 20 percent of the yearly demand for RDs.
- A one-time increase in investment in critical RD safety stock, equal to 40 percent of their yearly demand.

Figure 5.2
Reallocation of OA to RDs

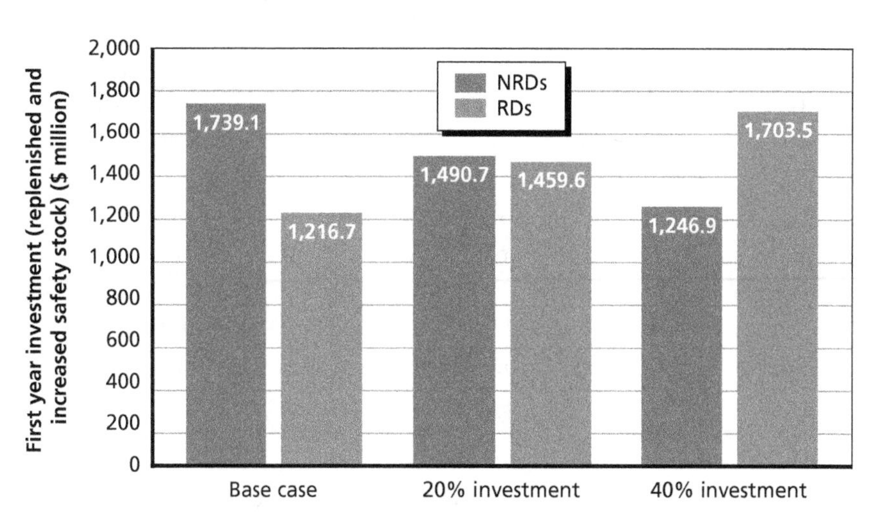

RAND *RR2496OSD-5.2*

In the latter two cases, funds for higher RD safety stocks would come from reducing investments in NRDs, in effect lowering their safety stock levels, with concomitant declines in their expected MA.

Figure 5.2 shows the OA investments used in the three cases. In the base case, most OA investment goes to NRDs, as they dominate demand and OA investment is based on replenishing that demand. The middle columns show a 20 percent redirection strategy, with OA investment even between the two populations, and the right-hand columns show the investments under the 40 percent approach, with a net increase of almost $500 million in RD OA, dedicated to increasing those items' safety levels.

Our simulation optimizes reallocation of OA to achieve maximum overall MA given historical variability of demand. Most of the investment goes to the more expensive items, primarily because most of the demand value comes from those NIINs. To further ensure that MA for the cheapest items is maximized, we modified the model to prevent reallocation of OA away from NRD NIINs with unit price below $100. This is done because inexpensive NIINs are have high MA, and reducing procurement of these items will likely have a significant reduction in MA while not improving the MA of RD NIINs. Reallocation funds from more expensive NRD NIINs to all RD NIINs will have the most increase of RD MA while limiting the impact of NRD MA. In general, then, because safety stock investment is allocated to dollar categories based upon the dollar-weighted demand of the categories, we see substantial reallocation of OA from expensive NRD NIINs to expensive RD items. Figures 5.3 and 5.4 show the dollar values moved by unit price group. Total OA is dominated by NIINs with unit price between $1,000 and $10,000, with RDs in that price range receiving

Figure 5.3
OA for NRDs by Unit Price Group and Investment Alternative

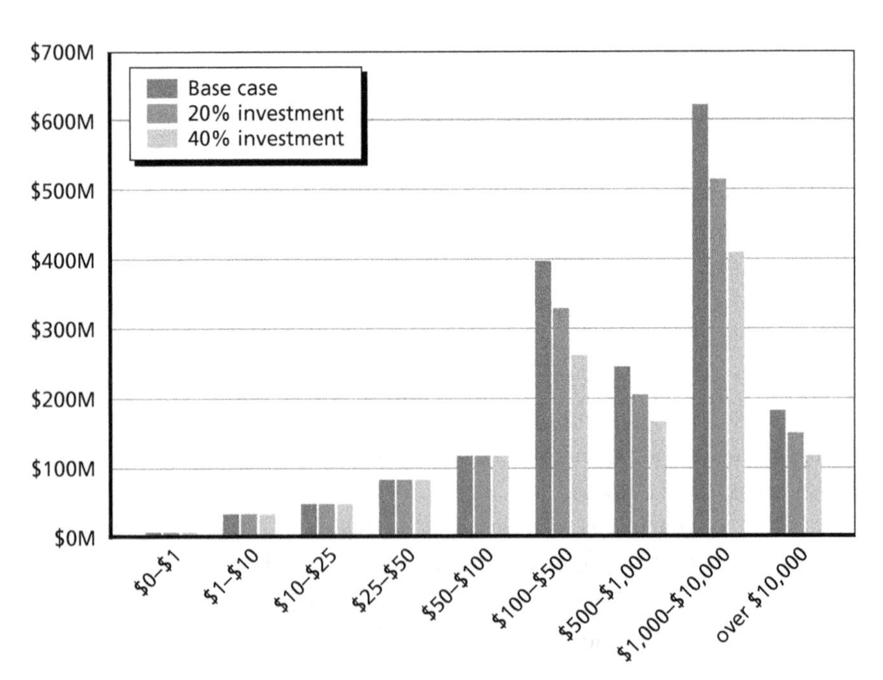

SOURCE: OA for NIINs with unit price below $100 protected.
RAND *RR2496OSD-5.3*

Figure 5.4
OA for RDs by Unit Price Group and Investment Alternative

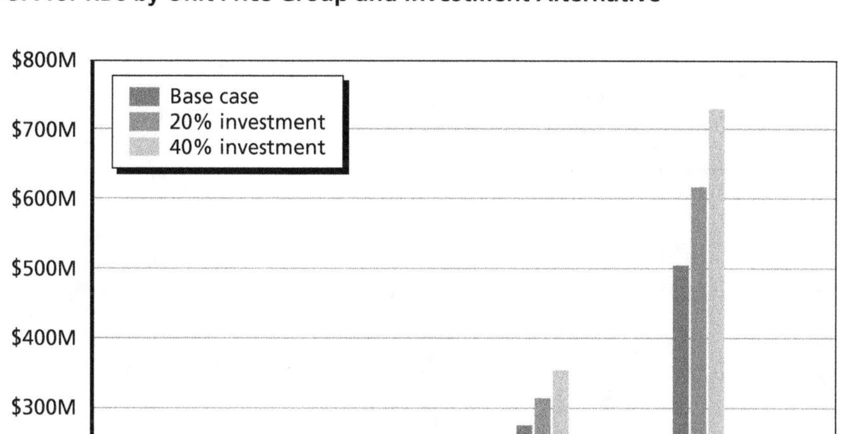

RAND *RR2496OSD-5.4*

the biggest increase in investment and the same group of noncritical NIINs showing the largest decrease in replenishment investment in the 20 percent and 40 percent cases.

Results

Figure 5.5 shows aggregate MA results for the three cases, with MA shown for the aggregate population and for RDs and the NRD population.

Overall simulated MA in the base case is 90 percent, matching DLA performance from 2014–2015. MA for the two subpopulations

Figure 5.5
Aggregate MA, by NIIN Type Across the Three Cases

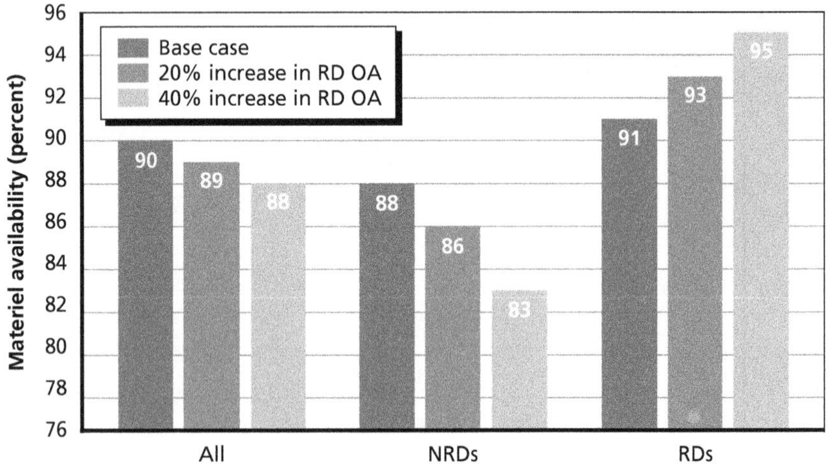

RAND *RR2496OSD-5.5*

is close: 88 percent for NRDs and 91 percent for RDs.[2] Reallocation of OA to RDs, especially the more expensive NIINs, increases MA for that population while reducing it for noncritical items. Because this investment strategy is not focused on aggregate optimized MA, overall MA drops. That is, MA for noncritical items sometimes drops more than MA for critical items increases. In the 20 percent reinvestment case, aggregate RD MA increases from 91 percent to 93 percent, while NRD MA drops from 88 percent to 86 percent. Overall MA drops from 90 percent to 89 percent. In the 40 percent reinvestment case, overall MA drops another percentage point overall compared to the 20

---

[2] RD MA is higher than that of NRD parts for two reasons: (1) the former group (perhaps surprisingly) includes a higher percentage of low-cost items than the latter, as 41 percent of demand RDs have unit prices below $10 compared to 29 percent NRDs; and (2) more of the NRD demand (25 percent) is for NIINs with Acquisition Advice Code Z than among RDs (only 5 percent). Acquisition Advice Code Z items, or insurance/numeric stockage objective items, are those which may be required occasionally or intermittently, to be stocked "in quantities no more than two minimum replacement units, except when document analysis supports a quantity that is more cost effective or is required to meet an explicit customer requirement" (U.S. Department of Defense, 2014).

Figure 5.6
RD MA Increases, by Unit Price Group

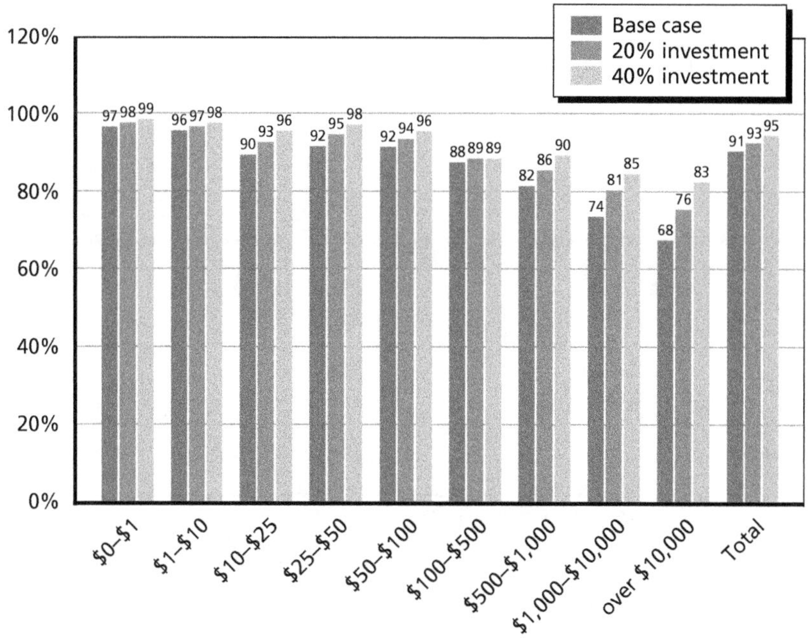

Unit price group	Base case	20% investment	40% investment
$0–$1	97	98	99
$1–$10	96	97	98
$10–$25	90	93	96
$25–$50	92	95	98
$50–$100	92	94	96
$100–$500	88	89	89
$500–$1,000	82	86	90
$1,000–$10,000	74	81	85
over $10,000	68	76	83
Total	91	93	95

RAND *RR2496OSD-5.6*

percent reinvestment case, increases another two points for RDs, and drops an additional three points for noncritical items.

The greatest increases in RD MA are in the higher-price items, which began at the lowest levels (see Figure 5.6).

Under the two alternative reinvestment approaches, total MA and NRD MA fall compared with the base case, while MA for RDs increases. Most of that increase goes to the more expensive RD parts, which have traditionally been relatively underresourced and thus have tended to have the lowest MA. Reallocating OA can yield substantial benefits for this group, while the MA for the more expensive NRDs is allowed to fall, as Figure 5.7 shows for the 20 percent reinvestment case.

Figure 5.7
Net Change in MA, by Price Group: Base Case, and 20 Percent Reinvestment Case

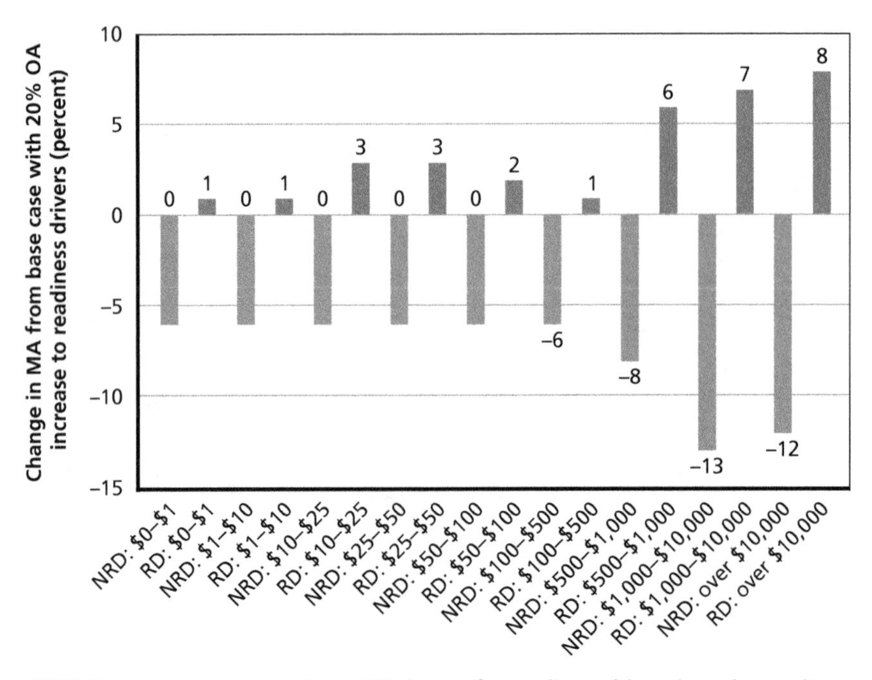

NOTE: Blue columns represent net MA change for readiness drivers in a given unit price group and red columns show net MA change for non-readiness drivers in a given unit price group. Absence of a column means there was zero net change in MA between the base case and the 20% reinvestment case.

RAND *RR2496OSD-5.7*

CHAPTER SIX

Toward an Integrated DoD Approach

Moving Toward a Revamped WSSP Using Critical Readiness Drivers

In 1994 DLA and the armed services rejected the idea that the number of items in the WSSP should be reduced, even as the program failed to deliver higher MA for WSSP items. Now the program is even larger—at least in terms of systems and parts included—and it continues to offer no meaningful support benefit for items in it deemed most critical; indeed, the parts deemed most critical account for almost two-thirds of everything DLA issues. But even if the WSSP *did* provide higher MA for high-priority WSSP NIINs, it is far from certain that the services would see a readiness benefit. As we show in this report, there is no necessary connection between a NIIN being in the WSSP (and having a high criticality score) and being an actual driver of weapon system readiness, as evidenced by field repairs.

Making the WSSP a program that successfully supports critical weapon system readiness, as argued here, requires two actions:

1. **The services should revise their method for determining critical RDs, with the almost certain implication that the list will be far smaller than that currently provided to DLA.** In other words, they must reverse the position they (along with DLA) took in 1994 that the group of participating NIINs should not be reduced. Concomitant with that, the services (and especially the Air Force) need to make their lists of critical weapon systems (and especially the highest-priority systems)

as stable as possible given operational requirements to avoid churn in NIIN management and ensure that critical NIINs can achieve and maintain high MA. The services will have to develop and defend their new methodologies to all stakeholders in the WSSP. Currently, there are wide differences among the number of high-priority WSSP NIINs offered by the services (see Table 2.7). This has not mattered greatly, as DLA has not made significant investment decisions based on those inputs, but if DLA changes its investment approach, those differences might matter a great deal. To avoid undermining the value of the program, all WSSP participants must make efforts to avoid even the appearance of "gaming the system" to inflate the number of parts needing increased investment. The methodologies used must be transparent, understandable, and defensible to all parties in the program.

2. **DLA must adopt a procurement strategy that prioritizes investment in these critical RD parts.** Assuming no new investment resources, that means that DLA (and the services) must accept that *lower-priority* items will necessarily see lower MA. It also means that DoD as a whole must accept "less efficient" MA results from an aggregate, enterprise-wide point of view. When enterprise-wide MA goals are set, reported on, and reacted to, investment will tend to flow to cheaper items that deliver more MA for the investment dollar. Targeting RDs (as reported here) would free up resources to allow more investment to more expensive over cheaper items. This will result in a lower enterprise-wide level of MA. However, it will yield a higher MA for those critical RDs. DLA will need to justify the level of performance it achieves internally, to the services, and to other interested parties and will likely have to differentiate reports on MA to highlight how it is achieving performance goals for targeted populations like RDs.

From Proof of Concept to an Implemented System

Developing agreed-upon methods for generating lists of critical RDs is key to improving the WSSP and building consensus for its outcomes. While the services are ultimately responsible for producing these new approaches, this report has presented examples of how that might be done and what challenges will be involved.

The key to the method presented here is to use field-level data reflecting the experience of flight line and motor pool mechanics responsible for achieving maximum readiness of unit equipment. These data need to be stored and archived in accessible databases, allowing for analysis of frequency and criticality and useful for building efficient critical RD NIIN lists. By "efficient" lists we mean that the underlying methodology must seek to identify the minimal number of critical RDs that provide the maximum impact on readiness. As discussed in Chapter Three, we sought out the smallest number of NIINs that made up at least 50 percent of all deadlining actions. In addition, each NIIN so identified had to contribute a sizable number of deadlining actions for the weapon system it was associated with. Lastly, deadlining actions had to exceed a minimum percentage of total requisitions made by the service for the particular NIIN.

Such a methodology already exists in a very mature form for Army ground systems and is currently used to prioritize stocking in brigade-level supply support activities (Girardini, Lackey, and Peltz, 2007). This methodology was enabled through the development of a new way of capturing critical deadlining part information into the EDA system (Peltz et al., 2002). (As noted, however, this is a very mature system for Army ground systems; a similar approach for aviation is still under development.)

The Air Force also maintains detailed histories of deadlining parts requests, or MICAPs, and historical MICAP information is archived in the LIMS–EV. While the Air Force does not yet use this information to stock at the base level, RAND applied logic to MICAP data similar to that used by the Army as part of this proof of concept.

With these two mature databases, and long-standing RAND experience in using them, the project team was able to develop a simulation model and execute the proof of concept analyses reported here.

Such a deep dive was not feasible under the constraints of our research agenda for the Navy and Marine Corps; similar work is yet to be done. A future list of Navy critical RDs will have to be especially pared down from its current WSSP population—which, as illustrated in Chapter Two, is especially large and dominates the entire DoD WSSP NIIN list (and indeed is a very large segment of everything DLA issues). The Navy's casualty report system may provide an effective way to categorize and rank critical RDs, along the lines presented here for Air Force and Army systems.[1] Work remains to be done for developing means for creating critical RD lists, although some work toward that goal has been done in the past.[2]

Developing and agreeing upon these new lists may take time, but progress toward improving the WSSP does not necessarily require complete and final development across all four services. Partial, phased implementation is possible. Because the Army methodology is the most mature, and the Air Force list could be compiled in a relatively short period of time, those services could submit new WSSP lists sooner rather than later, and these could be used to guide DLA investment decisions. That could be done as described in this report, where the investment base is the set of NIINs for which the two services dominate demand. Air Force and Army critical RD MA could be improved with no negative impact on Navy and Marine Corps demand. (Indeed, the same approach could be used for the Army independently if development of an Air Force list was delayed.) Ultimately, though, the new WSSP approach needs to be DoD-wide and must be based on critical RD NIIN lists (and the methodology underlying their selection) that will be seen as equitable, fair, and justifiable by all participants in the program.

---

[1] See, for example, "Casualty Report (CASREP) System" (undated).

[2] For historical background on earlier Marine Corps efforts to identify critical RDs, see Fricker and Robbins (2000, esp. Appendix A, "Critical Repairs and the ERO Fill Rate").

CHAPTER SEVEN

Summary, Conclusions, and Recommendations

Summary and Conclusions

The WSSP was established to provide guidance from the armed services to DLA for making investments aimed at prioritizing support for high-priority weapon systems. As has been documented in this report, the WSSP has not achieved that goal. The primary reason, as argued here, is the lack of the services' ability either to identify or communicate to DLA the relatively small sector of the repair part population that dominates readiness problems for their equipment. Instead, high-priority parts lists shared by the services with DLA tend to be inflated and, indeed, account for a large majority of *everything* DLA stores and issues. It is no surprise, then, that WSSP items, even those designated as being of high priority, are typically treated no differently from the run-of-the-mill materiel DLA manages, and generally show the same MA as anything else.

But the fact that the WSSP has not delivered higher MA for critical RDs does not mean that it is incapable of doing so. The Army already uses a critical RD approach to guide inventory stockage at its brigade-level supply support activities. The Air Force uses EXPRESS, the Execution and Prioritization of Repair Support System, to prioritize depot-level repair using MICAPs.[1] The same logic, this report argues, can and should be used to guide a new approach to supporting the WSSP. The report has attempted to make several crucial points:

---

[1] See Cook, Ausink, and Roll (2005).

- Identifying critical RD NIINs should be based on analysis of actual field-level maintenance data, and not on EC parameters typically assigned during weapon system acquisition (and not reexamined after fielding).
- To avoid creating parallel and confusing systems, the assignment of the EC should be modified over the course of the weapon system lifetime based on these field maintenance events.
- To maximize the value of the system, including any manual intervention, the list of critical RD NIINs should be limited, avoiding the inflation in number of NIINs submitted to the WSSP in order; for example, the critical list could be limited to the smallest number of NIINs that accounts for one-half of all deadlining actions. Ultimately, judgment will have to be used to develop the right level.
- Service critical RD NIIN lists need to be balanced enough to avoid overinvesting in one service's critical item list over those of other services; while current WSSP NIIN lists do show that imbalance, it has not mattered because DLA has not made investment decisions based on those lists; the goal, however, is to use the lists to make investments in the future, stressing the importance of ensuring that results are equitable.
- DLA should use these lists to guide decisions on investing OA in setting safety stock levels, and thus MA targets; this will necessarily result in less optimal aggregate MA as it will move OA investment from less critical and often cheaper items to more critical and often more expensive NIINs.
- As a result, overall MA for DLA-managed NIINs should decrease. This needs to be understood as a desirable outcome by all stakeholders, including the services, the Office of the Secretary of Defense (OSD), the Government Accountability Office, and Congress. DLA and OSD will need to be able to explain the approach and justify the results to external audiences.

Recommendations

RAND offers the following recommendations for improving the WSSP:

OSD Should Develop and Promulgate Policy Justifying and Providing for a New Approach to the WSSP

Such policy should seek to reduce the overall population of high-priority NIINs in the WSSP, with means to be determined by each service, with justification for doing so provided. It should also set the terms for balancing the size of those populations across the services in order to achieve fair allocation of resources across all services.

The Services Should Revise Their Approaches for Determining Critical RDs and Present Justifications for Those Approaches and the Resulting Lists

Using field maintenance data, as described in this report, is one way to identify critical RDs. While the ultimate determination will be made by the services, they need to be aware that equitable allocation of resources will imply constraints on how large those lists can be or what they imply in terms of OA investment.

The Services Should Seek to Make Minimal Changes in Their Assignment of System-Level Priority (or WSGCs) to Their Candidate Systems, and Should Review and Revise Their Critical RD NIIN Lists on a Regularly Scheduled Basis

Because achieving higher MA may have long lead times, and because there may be substantial sunk costs when making these investments, it is important that the services limit the volatility in their WSGC assignments and the resultant turbulence that would be created in their critical RD lists. Alternatively, however, the services must ensure that the lists of critical RD NIINs are as up to date as possible, this means both adding NIINs that are becoming increasingly important for readiness (e.g., due to aging) and removing those that are no longer critical.

DLA Should Change Policies and Procedures to Ensure Maximum Reasonable Investment of OA in Critical RD Safety Stock, Thereby Achieving Higher MA for Those Items

DLA should set target MA levels for the revised WSSP critical RD lists and allocate OA in order to achieve those targets, explicitly accepting lower MA for other items unless the overall budget is increased. Because aggregate MA is heavily influenced by the cheapest items (even if those items are not always the most important to driving readiness), DLA may need to establish explicit higher targets for more expensive NIINs, beyond what would be achieved through standard inventory management approaches.

DLA Should Report MA Results Disaggregated into RD and NRD Populations, as Well as Overall MA; After Advisement, OSD (and Other Stakeholders) Should Concur with an Approach That Lowers Overall MA in Exchange for Higher RD MA

OSD policy and required metrics, in particular, should reflect the revised investment strategy, and help support the disaggregated investment approach DLA would adopt under a new WSSP.

APPENDIX

The Simulation Model

The simulation is designed to support a proof of concept regarding the trade-off between RD and NRD investment allocation and MA. The simulation does not strive to replicate DLA's procurement logic, nor does is it intend to forecast actual MA as a function of investment allocation. The simulation's purpose is to explore the potential benefit from investment allocation trade-offs to help guide change in real-world strategy and policy.

Though we are not privy to the complete and highly complicated DLA replenishment logic, we quantify the empirical MA performance that the logic yields. In general, MA is very high for inexpensive NIINs and much lower for expensive NIINs, resulting in an overall unweighted MA of approximately 90 percent. The simulation is calibrated to mimic observed MA by unit price group and then used to model hypothetical investment scenarios.

There are three steps used to build and use the simulation for this purpose:

1. Create a computer simulation that models inventory, replenishment to DLA, and issues from DLA using observed DLA transaction data.
2. Calibrate the procurement logic in the simulation model to have the same MA results by unit price group, as displayed in Figures 4.2 and 4.3.
3. Simulate hypothetical changes in stock investment where increased investment in RD NIIN stock is decremented from NRD NIIN available procurement funds.

Once calibrated (i.e., step 2), we run hypothetical scenarios where available funds for NRD replenishment is reallocated and instead used to increase safety stock of RDs (step 3). This has the effect of increasing MA for the RDs and decreasing MA for the NRDs. We express the reallocation value as a percentage of the annual RD demand. Specifically, the following are the hypothetical scenarios that we simulate in which the dollar value relocated from NRD procurement funds to additional Readiness Driver RD safety stock is:

A. 20 percent of the annual RD demand value
B. 40 percent of the annual RD demand value.

For example, if RD annual demand value is $500 million, then $100 million is removed from the NRD OA and $100 million is used for additional RD safety stock in scenario B ($500 million × 20 percent = $100M).

Simulation Mechanics and Assumptions

The simulation is a stock-and-flow model that reflects replenishment logic, replenishment receipts, demands and issues, and stock levels.

Data Input and Preparation

The model uses the issues data for CY 2014 and CY 2015 from the DLA. Issues data are used instead of demand data because demand data have several complexities that go beyond the scope of the modeling. First, some of the demands are canceled for a variety of reasons that we did not deem necessary to model. Complexities include the following:

- customers no longer want certain items: back order durations are too long and the customers have obtained the parts through intraservice transfers
- if back orders exist, sometimes customers will submit multiple demands when, in fact, they only need one item
- some demands are partially filled given back orders and/or rationing due to low on-hand quantities.

All of these complexities are complicating factors that are difficult to model and can obscure the steady-state effects of investment strategies in this proof of concept. Therefore, we treat observed issues to customers as demands in the simulation model.

As input, we summarize the issues by week and sample the data, with replacement to be sampled again. This approach is similar to bootstrapping in statistics. The intent is not to simulate CY 2014 and CY 2015 issues in their exact sequence directly; the intent is to build a demand profile that is informed by CY 2014 and CY 2015 data but does not have the same sequence of demands. This is desirable in this proof of concept because we do not want the results to be unique to the particular CY 2014–CY 2015 sequence of demands but instead want the results to be based on a demand profile that resembles real-world data. Sampling with replacement provides that feature.

Step 1: Model Creation and Assumptions

The simulation models the inflows, outflows, and stock levels by NIIN and calculates performance metrics. When the stock levels are reviewed in the model, then the procurement logic decides how much to procure from vendors and submits a replenishment request. Given a PALT, the replenishment is observed and the stock is increased by the replenishment quantity. If stock is available and a demand event occurs in the simulation, the simulation will issue stock to customers, thus decreasing stock on hand. As the simulation time advances, the model keeps track of performance metrics, including the number of filled demands, due out quantities, demand quantities, and replenishment request quantities. These metrics allow postprocessing that calculates summary statistics about model performance, including MA by NIIN and unit price group.

Assumptions are necessary and are made regarding procurement logic and replenishment. We do not have visibility into DLA's procurement decisionmaking and logic and, therefore, must assume a reasonable replenishment logic to perform this proof of concept.[1] We assume

---

[1] If this effort were intended to be more than a proof of concept where actual performance changes were predicted based on procurement logic changes, then a thorough modeling of DLA's logic would be necessary.

an order-up-to policy with periodic review, a common ordering policy in practice and in inventory theory. In this policy the replenishment request quantity is the difference between an RO target and the current IP, where the IP is the sum of the quantity on hand and the difference of the due-in and due-out quantities (i.e., replenishment quantity = RO – [on hand + (due in – due out)]). We assume that the replenishment request quantity is decided each week in the simulation.[2]

The simulation time step is one week and the procurement logic is executed at the beginning of each week. A week was selected to make the model more tractable to run (as compared to daily or hourly time steps) and it seems reasonable to the research team that the inventory positions are reviewed at least once a week.

Each week the simulation would execute the following logic for each NIIN in each simulated week:

1. If the IP (on-hand quantity + due-in quantity – due-out quantity) is less than the RO, then it records a replenishment request for the difference between the RO and IP to arrive after a PALT lead time has passed in the simulation. Equation (X.1) shows inventory position calculation, and Equation (X.2) shows the order quantity, where order quantity is Q, the RO is R, inventory position is I, on-hand quantity is O, due-in quantity is D_I, and due-out quantity is D_O.[3]

$$I = O + D_I + D_O \tag{X.1}$$

$$Q = R - I = R - (O + D_I - D_O) \tag{X.2}$$

2. Increase the on-hand quantity by the quantity that was due in for the week.
3. Decrement the on-hand quantity by the quantity that was demanded in the week. The simulation allows the on-hand

---

[2] In actual practice, DLA considers replenishment decisions in real time depending on the OA, on-hand quantities, and priorities, among other attributes.

[3] If the order quantity calculation is negative, then no order is placed.

quantity value to be less than 0 in order to capture quantities due out.

4. Record the quantity that was immediately available and could be filled.

In the simulation, we make assumptions so that the model performance is that of a steady-state system rather than the transient short-term effects of the current system if investments are allocated. The goal is to demonstrate this proof of concept in the steady state and not as a function of current actual on-hand or vendor capabilities. The transient effects that are intentionally excluded are the effects on MA due to current on-hand quantity, current vendor on-hand quantity, production capabilities, and back order status. Therefore, we start the simulation on-hand quantity at the RO and progress in simulation time such that the inventory on-hand quantity reaches a steady state; then we start collecting performance statistics. This removes any transient effect of the on-hand starting quantity. Depending on the vendor on-hand quantity, production capabilities, and back order status, the PALT can and does change over time. Given that the vendors' effect is beyond the scope of this research and our desire is to not confound the vendors' effect on MA with the investment effect, in this proof of concept we assume a universal 26-week PALT. We then simulate the first 52 weeks and discard the performance so the simulated on-hand quantity can reach a steady state before performance statistics are captured.

Step 2: Calibrating the Model

Before running hypothetical investment scenarios, it is necessary to calibrate the model so that its performance reflects reality. As mentioned, we do not model DLA's full procurement logic because of its complexity. However, we can and do calibrate the model such that the assumed order-up-to policy has a simulated MA performance that mimics reality by unit price group. To calibrate, we adjust the ROs for each unit price group such that the simulated MA for each is approximately equivalent to the actual observed MA. This calibration is performed iteratively where simulated MAs are calculated and then ROs are adjusted repeatedly until the simulated MA is similar to that of the observed MA.

It is important to calibrate such that the starting point for the hypothetical scenarios at least has similar performance to reality. The hypothetical scenarios are naturally an extrapolation from the calibration but are defined to progressively extrapolate from the calibration to simulate the MA performance.

Calibration is conducted by adjusting the ROs so that the simulated MA is similar to that of the empirical MA. We do not do this at the NIIN level, but we adjust the ROs for all NIINs within a unit price group as a multiple of each NIIN's weekly demand standard deviation. As a starting point, we assume the demand has a theoretical Gamma probability distribution to calculate the percentile that corresponds to the target MA in each unit price group. We then add multiples of the weekly demand standard deviation and simulate so that the simulated MA is near the observed MA for the unit price group. Equation (X.3) shows how the RO starting point is estimated based on the theoretical Gamma probability distribution, where R_0 represents the starting point. In Equation (X.3), p is the target MA (e.g., 0.95 for unit price group <\$1) and the integral is the cumulative distribution of the Gamma distribution.[4] The *argmin* finds the cumulative distribution point (y) such that the cumulative distribution is equal to the target MA (p). In the integral, $\hat{\theta}$ and $\hat{k}$ are distribution parameter estimates from the sample weekly demand average ($\bar{x}$) and standard deviation ($\hat{s}$), where Equation (X.4) shows the relationships between the parameter estimates and the sample average and standard deviation.

$$R_0 = argmin_y \left| p - \int_0^y \frac{1}{\hat{\theta}^{\hat{k}} \Gamma\left(\hat{k}\right)} x^{\hat{k}-1} e^{-x/\hat{\theta}} \, dx \right| \tag{X.3}$$

$$\bar{x} = \hat{k}\hat{\theta} \text{ and } \hat{s}^2 = \hat{k}\hat{\theta}^2 \tag{X.4}$$

---

[4] Γ in Equation (X.3) is the gamma function, which is defined as $\Gamma(x) = \int_0^\infty t^{(x-1)} e^{(-t)} \, dt$.

R_0 is the starting point RO and is calculated for each NIIN. The multiples of the weekly standard deviation are added to the R_0, simulated against the sampled demand data, and a multiplication factor (m) that yields a simulated MA of near the target MA (p) is found. The factor m is different for each unit price group and is one factor that is applied universally across all NIINs within each unit price group. Equation (X.5) shows the multiples of the weekly standard deviation added to the R_0.

$$R = R_0 + m\hat{s} \tag{X.5}$$

Step 3: Modeling Hypothetical Investment Scenarios

In the hypothetical scenarios where investments are diverted from NRD available procurement funds to RD safety stock, we implement a logic that decrements the former by the latter's stock value. For this purpose, we assume that the available procurement funds in the simulation for the NRD NIINs are the annual observed issue value for those NIINs; in other words, we assume that the available procurement funds are exactly equal to the issue value in the CY 2014–CY 2015 data. To maintain high MA for inexpensive NRD NIINs, we do not decrement procurement funds for NIINs below $100. To do this, we calculate the RD investment value for each hypothetical scenario (i.e., 20 percent or 40 percent of the annual RD demand value). Then, for each unit price group, we calculate the NRD demand value as a proportion of total NRD demand value excluding those with unit price less than $100. That proportion is used to allocate the NRD procurement funds decrement. For example, in the Army and Air Force simulation 20 percent additional investment scenario, the total RD safety stock investment is $243 million. Additionally, the annual NRD procurement funds for unit price group $25–$50 is $84 million and is 5 percent of the total NRD annual procurement value. Then 5 percent of the $243 million will be decremented from the $25–$50 NRD procurement funds for net procurement funds of $72M (i.e., $84 million – 5 percent × $243 million; $84 million – $12 million = $72 million).

The decrement of NRD procurement funds does not leave enough funds to fully replace demand in the hypothetical scenarios. Therefore, we developed logic to allocate the available procurement funds to the NIINs that need the most procurement value within each unit price band each simulated week. This construct has the important attribute that every NIIN within each category will continue to be procured throughout the simulation because if a NIIN is not procured for a few weeks, it will rise in priority as demands continue and the desired procurement value continues to increase.

For example, suppose the desired procurement quantity is 5 for NIIN A, with a unit price of $10, and 2 for NIIN B, with a unit price of $20. Then the desired procurement value for NIIN A is $50 and NIIN B is $40. Therefore, NIIN A has a higher procurement priority according to this logic. Now suppose that there is only enough available procurement value to procure NIIN A for $50. In this case, NIIN A is procured and NIIN B is not procured in this simulated week. At this point, the desired procurement value is $0 for NIIN A (it was procured in this simulated week) and $40 for NIIN B (it was not procured in this week). In the next week, we see no demand for NIIN A, and two more demands for NIIN B, increasing their desired procurement to $80. At this time, NIIN B has a higher priority over NIIN A because NIIN B's desired procurement value is $80 and NIIN A's desired procurement value is $0. This approach prevents NIINs from continually being neglected each week.

References

"Casualty Report (CASREP) System," Integrated Publishing, undated. As of July 17, 2018:
http://www.tpub.com/gunners/274.htm

Cook, Cynthia R., John A. Ausink, and Charles Robert Roll, Jr., *Rethinking How the Air Force Views Sustainment Surge*, Santa Monica, Calif.: RAND Corporation, MG-372-AF, 2005. As of July 17, 2018:
https://www.rand.org/pubs/monographs/MG372.html

Defense Logistics Agency, *DLA Weapons Systems Support Program*, DLA Regulation 4140.38, Fort Belvoir, Va., June 9, 1989.

———, *DLA Retail Supply Chain Materiel Management Policy*, DLA Instruction 4140.08, Fort Belvoir, Va., March 11, 2015. As of July 17, 2018:
http://www.dla.mil/Portals/104/Documents/J5StrategicPlansPolicy/PublicIssuances/i4140.08.pdf

———, *Annual Financial Report: Fiscal Year (Unaudited) 2016*, Fort Belvoir, Va., 2016. As of August 21, 2017:
http://www.dla.mil/LinkClick.aspx?fileticket=GVJvIErDsDU%3D&portalid=104

Fricker, Ronald, and Marc Robbins, *Retooling for the Logistics Revolution: Designing Marine Corps Inventories to Support the Warfighter*, Santa Monica, Calif.: RAND Corporation, MR-1096-USMC, 2000. As of July 17, 2018:
https://www.rand.org/pubs/monograph_reports/MR1096.html

Girardini, Kenneth, Arthur W. Lackey, and Eric Peltz, "Stockage Determination Made Easy," Santa Monica, Calif.: RAND Corporation, RP-1272. Reprinted from *Army Logistician*, Vol. 39, No. 4, July–August 2007, pp. 12–15. As of July 17, 2018:
https://www.rand.org/pubs/reprints/RP1272.html

Hanks, Chris, *How DLA's Supply Performance Affects Air Force Readiness*, Bethesda, Md.: Logistics Management Institute, Report DL901R1, October 1990. As of July 17, 2018:
http://www.dtic.mil/dtic/tr/fulltext/u2/a230291.pdf

Headquarters, U.S. Department of the Army, *Army Participation in the Defense Logistics Agency Weapon System Support Program*, Army Regulation 711-6, Washington, D.C.: U.S. Department of the Army, May 15, 2009. As of July 17, 2018:
http://usahec.contentdm.oclc.org/cdm/ref/collection/p16635coll11/id/2678

———, *Army Aviation Maintenance*, TC 3-04.7 (FM 3-04.500), February 2, 2010. As of August 21, 2017:
http://www.apd.army.mil/epubs/DR_pubs/DR_a/pdf/web/tc3_04x7.pdf

Office of the Department of Defense Inspector General, *Defense Logistics Agency's Weapons Systems Support Program*, Report No. 95-027, Alexandria, Va., November 9, 1994. As of July 17, 2018:
https://media.defense.gov/1994/Nov/09/2001714960/-1/-1/1/95-027.pdf

———, *Audit Report on Mission Essentiality Coding*, Report No. 97-086, Alexandria, Va., February 3, 1997. As of July 17, 2018:
https://media.defense.gov/1997/Feb/03/2001715387/-1/-1/1/97-086.pdf

Peltz, Eric, Patricia Boren, Marc Robbins, and Melvin Wolff, *Diagnosing the Army's Equipment Readiness: The Equipment Downtime Analyzer*, Santa Monica, Calif.: RAND Corporation, MR-1481-A, 2002. As of August 21, 2017:
http://www.rand.org/pubs/monograph_reports/MR1481.html

Peltz, Eric, Amy G. Cox, Edward W. Chan, George E. Hart, Daniel Sommerhauser, Caitlin Hawkins, and Kathryn Connor, *Improving DLA Supply Chain Agility: Lead Times, Order Quantities, and Information Flow*, Santa Monica, Calif.: RAND Corporation, RR-822-OSD, 2015. As of July 17, 2018:
https://www.rand.org/pubs/research_reports/RR822.html

Petcoff, Russell P., "Air Force Program Recognized for Excellence in Government," U.S. Air Force, May 5, 2010. As of August 21, 2017:
http://www.af.mil/News/Article-Display/Article/116767/
air-force-program-recognized-for-excellence-in-government/

Robinson, Nathaniel, *The Defense Logistics Agency: Providing Logistics Support Throughout the Department of Defense*, Maxwell Air Force Base, Ala.: Air University Press, Research Report No. AU-ARI-92-4, October 1993. As of August 21, 2017:
http://www.dtic.mil/dtic/tr/fulltext/u2/a273791.pdf

U.S. Department of the Air Force, *Materiel Management Policy,* Air Force Instruction 23-101, Washington D.C.: U.S. Department of the Air Force, 2016. As of August 14, 2018:
http://static.e-publishing.af.mil/production/1/af_a4/publication/afi23-101/
afi23-101.pdf

U.S. Department of Defense, *DoD Supply Chain Management Procedures: Demand and Supply Planning*, Manual 4140.01, Vol. 2, Washington, D.C.: U.S. Department of Defense, February 10, 2014. As of August 21, 2017: http://www.esd.whs.mil/Portals/54/Documents/DD/issuances/414001m/414001m_vol02.pdf

U.S. Department of the Navy, Headquarters, U.S. Marine Corps, *Supply Chain Integration; Marine Corps Participation in the Defense Logistics Agency (DLA) Weapon System Support Program (WSSP)*, Marine Corps Bulletin 4105, Washington D.C.: U.S. Department of the Navy, December 1, 2011. As of August 21, 2017: http://www.marines.mil/Portals/59/Publications/MCBUL 4105.pdf

Printed in the USA
CPSIA information can be obtained
at www.ICGtesting.com
CBHW061114101223
2545CB00009B/881